# Naval Diesel Engineering
## The Fundamentals of Operation, Performance and Efficiency

## Onturo D. Johnson

authorHOUSE®

*AuthorHouse™*
*1663 Liberty Drive*
*Bloomington, IN 47403*
*www.authorhouse.com*
*Phone: 833-262-8899*

*Published by AuthorHouse  04/07/2022*

*ISBN: 978-1-6655-5606-4 (sc)*
*ISBN: 978-1-6655-5614-9 (e)*

*Print information available on the last page.*

*This book is printed on acid-free paper.*

# ACKNOWLEDGEMENT

The information contained herein has been adapted from the *Engineman 1 & C* and *Engineman 3 & 2*, prepared by the Bureau of Naval Personnel, NAVPERS 10543, First edition 1954 and NAVEDTRA 14331, First edition 2000, 2003, respectively: US Government Printing Office Washington, DC. 20402. The majority of this text was prepared by the Training Publications Center, Naval Personnel Program Support Activity, Washington, D.C. To the extent, this book may contain text in the public domain; the Author makes no claim of ownership. The Author is credited for text compilation and editing. United States Navy photographs taken by MC2 Dominique A Pinero, Cpl. Theodore W. Ritchie and released to the public.

# CONTENTS

# INTRODUCTION

*Naval Diesel Engineering: Fundamentals of Operation, Performance and Efficiency* introduces the common types of fuels and the hardware systems that store, clean, transfer, and finally inject the fuel into the engine for burning, general types of installations while recognizing the fundamental starting, operating, and stopping procedures for a diesel engine under normal operating conditions aboard naval ships. The prime concern of Navy Diesel Engineers (or Engineman) is to keep the machinery for which they are responsible operating in the most efficient manner possible. Additionally, they know that engine efficiency and performance depend upon much more than just operating the throttle and changing oil at prescribed intervals. In troubleshooting an internal combustion engine, whether diesel or gasoline, this book will cover similar procedures.

You should be able to identify the characteristics of engine fuels, diesel engine fuel systems, describe unit fuel injector systems, understand how the operating speed of a diesel engine is controlled, and identify methods used to purge air from the fuel system of a diesel engine. Official illustrations and details of overhaul, maintenance, and repair have been purposely omitted from this text. The text focuses on diesel engine system components and control devices and discusses diesel engine trouble shooting and best practices for engineering professionals seeking optimal engine efficiency. This book describes some of the main causes for diesel engines failing to start, failing to maintain power after starting, overheating, and having abnormal exhaust.

Emphasis is placed on the various types of troubles that arise in connection with this machinery, the causes of these troubles, the symptoms by which they are indicated, and the remedial procedures to be followed from the Navy Diesel Engineers perspective.

# CHAPTER 1

# DIESEL FUEL SYSTEMS AND ENGINE CONTROL DEVICES

In the first part of this chapter, we will discuss the common types of fuels and the hardware systems aboard naval ships that store, clean, transfer, and inject the fuel into the engine for burning and how injected fuel maintain the engine at a set speed.

After reviewing the information in this chapter, you should be able to identify the characteristics of engine fuels in terms of properties, combustion, volatility, and turbulence. You should also be able to identify diesel engine fuel systems in terms of external fuel systems and fuel injection systems. You should be able to describe the jerk-type, distributor-type, and unit fuel injector systems in terms of design and function of components and methods of operation. Additionally, you should understand how the operating speed of a diesel engine is controlled. Finally, you should be able to identify methods you can use to purge air from the fuel system of a diesel engine.

## Diesel Engine Fuel Requirements

The fuels burned in the internal-combustion engines used by the Navy must meet the specifications prescribed by the Naval Sea Systems Command. Therefore, the problem of selecting a fuel with the required properties is not your responsibility. The Navy Diesel Engineer's primary responsibility is to follow the rules and regulations dealing with the proper use of fuels. He must strictly adhere to all prescribed safety precautions. He must also take every possible precaution to keep fuel as free as possible from impurities. Even though proper handling and use are his prime responsibilities with respect to fuel, knowing the characteristics of fuels will help him understand some of the problems in engine operation and maintenance. At the time of manufacture, fuels are generally clean and free from impurities.

However, the processes of transferring, storing, and handling fuel tend to increase the danger of contamination with foreign materials, a condition that can interfere with engine performance. Sediment and water in fuel can cause engine wear, gumming, and corrosion in the fuel system. Foreign materials in fuel can also cause an engine to operate erratically with a loss in power. For these reasons, periodic inspection, cleaning, and maintenance of fuel handling and filtering equipment are necessary. Because of the differences in the combustion processes and in the fuel systems of diesel and gasoline engines, the fuels for these engines must be refined to meet different requirements. In general, diesel engines require a particularly clean fuel; otherwise, the closely fitted parts of the injection equipment will wear rapidly and the small passages that create the fuel spray within the cylinders will become clogged. The diesel fuel must have a composition that permits its injection into the cylinders in a fine mist or fog. As injected diesel fuel enter the cylinder, it must have ignition qualities that permit the fuel to ignite properly and burn rapidly.

**Volatility and Engine Operation**

The ability of a liquid to change to vapor is known as *volatility*. All liquids tend to vaporize at atmospheric temperatures, but their rates of vaporization vary. The rate of vaporization increases as the temperature increases and as the pressure decreases (temperature is more important than pressure). In general, for a given temperature, a highly volatile fuel will vaporize more readily and at a faster rate than a fuel with a lower volatility.

The volatility of fuel affects engine-starting, length of warmup period, fuel distribution, and engine performance. When compared to diesel fuel (F-76), gasoline is much more volatile. High volatility, however, can also result in fuel dilution of the lube oil in the crankcase.

## Injection, Ignition, and Combustion

The self-ignition point of a fuel is a function of temperature, pressure, and time. In a properly operating diesel engine, the intake air is compressed to a high pressure (increases the temperature), and the injection of fuel starts a few degrees before the piston reaches top dead center (TDC). The fuel is ignited by the heat of compression shortly after fuel injection starts and combustion continues throughout the injection period. Combustion in a diesel engine is much slower than it is in a gasoline engine, and the rate of pressure rise is relatively small. Immediately after injection, the atomized fuel partially evaporates with a resultant chilling of the air in the immediate vicinity of each fuel particle. However, the extreme heat of compression rapidly heats and vaporizes the fuel droplets to the self-ignition point and combustion begins. The fuel particles burn as they mix with the air. The smaller particles burn rapidly, but the larger particles take more time to ignite as heat bring them to the self-ignition point.

There is a time delay between the period when fuel is first delivered into the cylinder and when it reaches it self-ignition point. This delay is referred to as *ignition delay* or *lag*. The duration of the ignition delay is dependent upon the characteristics of the fuel, the temperature and pressure of the compressed air in the combustion space, the average size of the fuel particles, and the amount of turbulence present in the space. During this stage of combustion, the temperature and pressure within the space rise rapidly, this reduces the ignition delay in the fuel particles injected later in the combustion process versus those injected earlier.

In a diesel engine, the delay period between the start of injection and the start of self-ignition is sometimes referred to as the first phase of combustion. The second phase of combustion is ignition of the fuel injected during the first phase and the rapid spread of the flame through the combustion space, as injection continues. The resulting increase in temperature and pressure reduce the ignition lag for the fuel particles entering the combustion space during the remainder of the injection period. Remember, only a portion of the fuel has been injected during the first and second phases. As the remainder of the fuel is injected, the third or final phase of combustion takes place. The increase in temperature and pressure during the second phase is sufficient to cause most of the remaining fuel particles to ignite with practically no delay in the third phase as they come from the injection equipment. The rapid burning during the final phase of combustion causes an additional, rapid increase in pressure.

The knock that occurs during the normal operation of a diesel engine should not be confused with detonation. Generally, *detonation* in a diesel engine is caused by a simultaneous combustion of all particles of the fuel spray in the cylinder. *Combustion (Diesel) Knock* in a diesel engine is directly related to the amount of ignition delay and will take place at the end of the second phase. Diesel knock occurs from the rapid burning of large amounts of fuel (gathered in the cylinder before combustion begins). Whether combustion is normal or whether detonation occurs is determined by the amount of fuel that is ignited instantaneously. The greater the amount of fuel that ignites at one time, the greater the pressure rise and the more severe the knock. Detonation in a diesel engine is generally caused by too much delay in ignition. The greater the delay, the greater the amount of fuel that accumulates in the cylinder before ignition. When the ignition point of the excess fuel is reached, all of this fuel ignites simultaneously, causing extremely high pressures in the cylinder and an undesirable knock. Consequently, detonation in a diesel generally occurs at what is normally considered the start of the second phase of combustion. Detonation in a diesel may occur when the engine is not warmed up sufficiently or when fuel injection equipment is not operating properly. These conditions may allow excessive fuel to accumulate in the cylinder.

Even though diesel fuel must have the ability to resist detonation, it must ignite spontaneously at the proper time under the pressure and temperature conditions existing in the cylinder. The ease with which a diesel fuel ignites and the manner in which it burns determines the ignition quality of the fuel. The ignition quality of a diesel fuel is determined by its *cetane rating* or *cetane number* In fact, the cetane rating of a diesel fuel is identified by its cetane number. The higher the cetane number, the less lag there is between the time the fuel enters the cylinder and the time it begins to burn. The cetane number of a diesel fuel is derived from a comparison test. The cetane number of diesel fuel is the numerical result of an engine test designed to evaluate fuel ignition delay. To establish the cetane number scale, two reference fuels are used, cetane and heptamethylnonane. Cetane has an excellent ignition quality (100), and heptamethylnonane has a very poor ignition quality (15). The cetane rating of a fuel in which the ignition quality is unknown can be determined by a comparison of the performance of the fuel with that of a reference fuel. The cetane number represents the percentage of pure cetane in a reference fuel that will just match the ignition quality of the fuel being tested. A higher cetane number means a quicker burning of the fuel, a condition that tends to result in easier engine starting, particularly in cold weather.

**Turbulence and Combustion in Diesel Engines**

In both gasoline and diesel engines, the fuel and air must be properly mixed to obtain efficient combustion. In gasoline engines, mixing of the fuel and air takes place outside the cylinder. Depending upon the design of the system, mixing will occur in one of two places: (1) within the carburetor in the carburetor-type system or (2) at the intake ports in the fuel injection-type systems. In both designs, the proper mixture is forced into the cylinder to be compressed. In the diesel engine, however, fuel in the form of small particles is sprayed into the cylinder after the air has been compressed; therefore mixing takes place within the cylinder. If each particle of fuel is to be surrounded by sufficient air to burn it completely (that is, if proper air-fuel mixture is to be obtained), the air in the combustion space must be in motion. This air motion is called *turbulence*. Various means are used to create turbulence. Design of engine equipment and parts and, in some engines, a process called *precombustion* enter into the creation of proper turbulence within the cylinder of an engine.

### Methods of Creating Turbulence

Fuel is distributed in the cylinders of a diesel engine by injection nozzles, which atomize the fuel and direct it to the desired portions of the combustion space. *Fuel Injection* creates some turbulence, but not enough for efficient combustion. In 2-stroke cycle engines, scavenging-air PORTS are designed and located so that the intake air enters the cylinder with a whirling or circular movement. The movement of the air continues through the compression event and aids in mixing the air and fuel when injection occurs. While fuel injection and the ports in 2-stroke cycle engines aid in creating air movement, additional turbulence is created in most engines by special shapes in the combustion space. These shapes may include the piston crown and that portion of the cylinder head that forms part of the main combustion space. In some engines, auxiliary combustion chambers are provided as part of the combustion space to aid in mixing the fuel and air.

Even though there are many types of combustion chambers, all are designed to produce one effect-to bring sufficient air in contact with the injected fuel particles to provide complete combustion at a constant rate. Combustion chambers may be broadly classified under four types: open, precombustion, turbulence, and divided chamber. The last three terms are more commonly used to identify auxiliary combustion chambers; all are associated with the process called *precombustion.* Of the three types of chambers, the open combustion chamber is the simplest in design. The fuel is injected directly into the top of the combustion space. The piston crown and (in some designs) the cylinder head are shaped to cause a swirling motion of the air as the piston moves toward TDC during the compression event. There are no special chambers to aid in creating turbulence. Open combustion chambers require higher injection pressures and a greater degree of atomization than other types to obtain the same degree of turbulence and mixing.

**Precombustion and Turbulence**

Some diesel engines have an auxiliary space or chamber at or near the top of each main combustion space. These chambers receive all or part of the injection fuel and condition it for final combustion in the main combustion chamber of the cylinder. This conditioning, called precombustion, involves a partial burning of the fuel before it enters the main combustion space. Precombustion helps to create the turbulence needed for the fuel and air to be properly mixed. Because of differences in designs, the manner in which precombustion aids in creating turbulence differs from one type of auxiliary combustion chamber to another. For this reason, we will discuss three types of auxiliary chambers by their common names—PRECOMBUSTION CHAMBERS, TURBULENCE CHAMBERS, and AIR or ENERGY CELLS.

The precombustion chamber, spherical in shape, is located in the cylinder head directly over the center of the piston crown. The precombustion chamber is connected to the main combustion space of the cylinder by a multiple orifice called a burner. During the compression event, a relatively small volume of compression-heated air is forced through the burner into the precombustion chamber. Heat stored by the burner increases the temperature of the compressed air and facilitates initial ignition.

Fuel is atomized and sprayed into the hot air in the precombustion chamber and combustion begins. Only a small part of the fuel is burned in the precombustion chamber because of the limited amount of oxygen. The fuel that does burn in the chamber creates enough heat and pressure to force the fuel, as injection continues, into the cylinder at great velocity. The velocity of the fuel entering the main combustion space and the shape of the piston crown help to create the necessary turbulence within the cylinder.

Engines that have precombustion chambers do not require fuel injection pressures as great as engines that have open-type chambers. In addition, the spray of injected fuel can be coarser, since the precombustion chamber functions to atomize the fuel further before the fuel enters the cylinder.

Some engines have auxiliary combustion chambers. They differ from precombustion chambers in that nearly all of the air supplied to the cylinder during the intake event is forced into the auxiliary chamber during the compression event. Auxiliary chambers in which this occurs are sometimes referred to as TURBULENCE CHAMBERS, which there are several variations of them.

Turbulence is created in the auxiliary chamber as compression, injection, and combustion take place. In engines with turbulence chambers, there is very little clearance between the top of the piston and the head when the piston reaches TDC. For this reason, a high percentage of the air in the cylinder is forced into the turbulence chamber during the compression event. The shape of the chamber (usually spherical) and the size of the opening through which the air must pass help to create turbulence. The opening to the turbulence chamber becomes smaller as the piston reaches TDC, thereby increasing the velocity of the air. Velocity plus deflection of the air as it enters the auxiliary chamber creates considerable turbulence. Fuel injection is timed to occur when the turbulence in the chamber is the greatest. This ensures a thorough mixing of the air and fuel. The greater part of combustion takes place within the turbulence chamber and is completed as the burning gases expand and force the piston down in the power event.

In some high-speed diesel engines, turbulence is created by an auxiliary chamber referred to as an ENERGY (AIR) CELL. Energy cells differ in design and location. In most engines, the cells are located in the cylinder heads. One type of energy cell that is located in the cylinder head is a divided chamber and turbulence chamber. The Lanova cell is the divided chamber type. There is also a divided auxiliary combustion chamber and we will discuss how the energy cell system works. The Lanova design employs a combustion chamber consisting of two rounded spaces cast in the cylinder head. The inlet and exhaust valves open into the main combustion chamber. The fuel-injection nozzle lies horizontally, pointing across the narrow section where the lobes join. Opposite the nozzle is the two-part energy cell, which contains less than 20 percent of the main-chamber volume.

The action is as follows: During the compression stroke, the piston forces air into the energy cell. Near the end of the stroke, the nozzle sprays fuel across the main chamber in the direction of the mouth of the energy cell. While the fuel charge is traveling across the center of the main chamber, between a third and a half of the fuel mixes with the hot air and burns at once. The remainder of the fuel enters the energy cell and starts to burn there, being ignited from the fuel already burning in the main chamber.

At this point, the cell pressure rises sharply, causing the products of combustion to flow at high velocity back into the main combustion space. This sets up a rapid swirling movement of fuel and air in each lobe of the main chamber, promoting the final fuel-air mixing and ensuring complete combustion. The two restricted openings of the energy cell control the time and rate of expulsion of the turbulence-creating blast from the energy cell into the main combustion space.

Therefore, the rate of pressure rise on the piston is gradual, resulting in smooth engine operation.

The divided combustion chamber is similar, in some respects, to other types of chambers. It is similar to an open combustion chamber in that the main volume of air remains in the main combustion chamber and principal combustion takes place there. Both the divided chamber and the turbulence chamber depend on a high degree of turbulence to ensure thorough mixing and distribution of the fuel and air. However, turbulence in a divided combustion chamber is dependent on thermal expansion caused by combustion in the energy cell and not on engine speed as in other types of auxiliary combustion chambers.

**Naval Distillate Diesel Fuel**

The fuel normally used in diesel engines is naval distillate (NATO symbol F-76), but other fuels such as JP-5 (NATO symbol F-44) and naval distillate lower pour point (NATO symbol F-75) are also used. Code F-76 and F-75 fuels are compatible and can be mixed in all proportions. Code F-44 and F-75 fuels are authorized for use in diesel engines where there is a logistic advantage for use. At present, most ships carry naval distillate fuel (F-76) for boilers and for diesel engines.

**Shipboard Fuel Testing**

Normally, fuel is procured through the military supply system. It is reasonable to assume that fuel received from the military supply system will meet the requirements of the applicable military fuel specification because of the extensive quality surveillance procedures used by this system. However, when delivered to ships by fleet oilers, fuel can be contaminated by solids and water. Solids and water must be removed by the receiving ship through settling and stripping or by use of purification equipment.

The fuel testing equipment items described in this section will helps diesel engineers identify the solids and water that must be removed by stripping and determine whether the purification equipment is functioning adequately to remove the solids and the water.

Required shipboard fuel testing equipment items for all ships are as follows:

| Name of Test | Equipment (methods) |
|---|---|
| Visual | Glass sample bottle |
| Bottom sediment and water (BS&W) | Laboratory centrifuge |
| Flashpoint | Pensky-Martens closedcup tester |
| API gravity | Hydrometer range: 29-41 and 39-51 |

In addition to these four tests, if diesel engineers are assigned to a ship with gas turbine propulsion plants or helicopter in-flight refueling (HIFR) capability, they will be required to have the following additional testing equipment:

1. A.E.L. free water detector Mk II

2. A.E.L. contaminated fuel detector Mk III

**Fuel Systems**

In diesel engine construction, two types of fuel systems are used: external fuel systems and fuel injection systems.

**External Fuel Systems**

The fuel system in a diesel-powered naval vessel must be installed, operated, and maintained with the same care and supervision as the ship's engines. Inspections, maintenance, and operation of fuel tanks and fuel-handling equipment must be carried out according to the

Ships' 3-M Systems and the Naval Ships' Technical Manual, chapters 541 and 542.

The fuel is pumped from the storage tank to the service or day tank, and from there it is delivered to the fuel injection equipment on the engine. It is good practice to remove sediment and water from the fuel before it enters the service tank. This is usually done with a centrifugal purifier. The fuel is transferred from the service tank by an engine-driven pump (also called a booster, transfer, or primary pump) through a metal-edge strainer and a cartridge-type replaceable element filter to a header at the inlet of the fuel injection equipment installed on the engine. Excess fuel (that is NOT used for combustion) is returned to the service tank.

The service or day tank is usually vented to the atmosphere and mounted at a high point in the fuel system, which allows the weight of the fuel to pressurize the external system and prevent air from leaking into the system. The presence of air in the fuel will interfere with proper operation of the fuel injection equipment.

The engine-driven fuel transfer pump is the positive-displacement type. Usually the fuel transfer pump is equipped with a built-in relief valve to ensure constant pressure to the injection equipment. The fuel strainer is located on the suction side of the transfer pump, and the filter is connected into the system on the discharge side of the pump. The pressure drop across these filters and strainers increases with time. For this reason, some systems have a relief valve in the line before the cartridge filter so that the bypassed fuel can be returned directly to the supply tank. The relief valve provides a more constant fuel supply pressure.

The primary metal-edge fuel strainer used in Navy installations is a duplex type, which is actually two complete strainers connected by a suitable header or piping. This arrangement allows either strainer to be completely cut out of the system for cleaning or repair while all of the fuel flows through the other strainer. The fuel flows from the outside to the inside. In Navy-approved strainers, the spaces between the leaves, or ribbons, which act as fuel passages, are between 0.001 and 0.0025 inch. The pressure drop across these strainers must not be allowed to exceed 1.5 psi when a fuel flow is equal to the full capacity of the fuel pump. In some engines, a duplex strainer is placed between the fuel supply tank and the transfer pump and, during operation, may be working under a vacuum.

The secondary, cartridge-type, fuel filter contains elements that must conform to Navy specifications. The pressure drop across clean and new elements should not be allowed to exceed 4.5 psi. The elements should be changed when the pressure drop reaches the value specified in the manufacturer's instruction book. The sump of the filters and strainers should be drained as often as practicable, preferably when the fuel is flowing.

## Centrifugal Purifiers

Detailed instructions are furnished with each purifier concerning its construction, operation, and maintenance. When you are responsible for the operation and maintenance of a purifier, study the appropriate NAVSEA technical manual and follow the instructions carefully. The following sections will provide general information on the methods of purification and the principles of operation of centrifugal purifiers. Centrifugal purifiers are used for purification of both fuel and lubricating oil. A purifier may remove both water and sediment, or it may remove sediment only. When water is involved in the purification process, the purifier is usually called a *separator*. When the principal contaminant is dirt or sediment, the purifier is used as a *clarifier*. Purifiers are generally used as separators for the purification of fuel. When used for purification of a lubricating oil, a purifier may be used as either a separator or a clarifier. Whether a purifier is used as a separator or a clarifier depends on the water content of the oil that is being purified.

NOTE: According to the Naval Sea Systems Command (NAVSEA) at the time of the writing of this training manual, purifiers are not authorized for use with diesel engine lubricating oil systems. The following general information will help you understand the purification process, the purposes and principles of purifier operation, and the basic types of centrifugal purifiers in use in naval service.

**Principles of Operation**

In the purification of fuel, centrifugal force is the fundamental principle of operation. Centrifugal force is force that is exerted upon a body or substance by rotation. Centrifugal force impels the body or substance outward from the axis of rotation.

A centrifugal purifier is essentially a container, which is rotated at high speed while contaminated fuel is forced through, and rotates with, the container. However, only materials that are insoluble in the fuel can be separated by centrifugal force. For example, JP-5 or naval distillate cannot be separated from lubricating oil, nor can salt be removed from seawater by centrifugal force. Water, however, can be separated from fuel because water and fuel do not form a true solution when they are mixed. Furthermore, there must be a difference in the specific gravities of the materials before they can be separated by centrifugal force.

When a mixture of fuel, water, and sediment stands undisturbed, gravity tends to form an upper layer of fuel, an intermediate layer of water, and a lower layer of sediment. The layers form because of the specific gravities of the materials in the mixture. If the fuel, water, and sediment are placed in a container, which is revolving rapidly around a vertical axis, the effect of gravity is negligible in comparison with that of the centrifugal force. Since centrifugal force acts at right angles to the axis of rotation of the container, the sediment with its greater specific gravity assumes the outermost position, forming a layer on the inner surface of the container. Water, being heavier than fuel, forms an intermediate layer between the layer of sediment and the fuel, which forms the innermost layer. The separated water is discharged as waste, and the fuel is discharged to the service tank or the day tank. The solids remain in the rotating unit.

The size of the particles, the viscosity of the fluids, and the time during which the materials are subjected to the centrifugal force further affect separation by centrifugal force.

In general, the greater the difference in specific gravity between the substances to be separated and the lower the viscosity of the fuel, the greater will be the rate of separation.

**Types of Centrifugal Purifiers**

Two basic types of purifiers are used in Navy installations. Both types use centrifugal force. Principal differences in these two machines exist, however, in the design of the equipment and the operating speed of the rotating elements. In one type, the rotating element is a bowl-like container that encases a stack of discs. This is the disc-type *DeLaval* purifier, which has a bowl operating speed of about 7,200 rpm. In the other type, the rotating element is a hollow cylinder. This machine is the tubular-type *Sharples* purifier, which has an operating speed of 15,000 rpm.

**Disc-Type Purifier**

The bowl is mounted on the upper end of the vertical bowl spindle, which is driven by means of a worm wheel and friction clutch assembly. A radial thrust bearing at the lower end of the bowl spindle carries the weight of the bowl spindle and absorbs any thrust created by the driving action. Contaminated fuel enters the top of the revolving bowl through the regulating tube. The fuel then passes down the inside of the tubular shaft, out the bottom, and up into the stack of discs. As the dirty fuel flows up through the distribution holes in the discs, the high centrifugal force exerted by the revolving bowl causes the dirt, sludge, and water to move outward. The purified fuel is forced inward and upward, discharging from the neck of the top disc. The water forms a seal between the top disc and the bowl top. (The top disc is the dividing line between the water and the fuel.) The discs divide the space within the bowl into many separate narrow passages or spaces. The liquid confined within each passage is restricted so that it can flow only along that passage. This arrangement minimizes agitation of the liquid as it passes through the bowl. It also forms shallow settling distances between the discs.

Any water, along with some dirt and sludge, separated from the fuel, is discharged through the discharge ring at the top of the bowl. However, most of the dirt and sludge remains in the bowl and collects in a more or less uniform layer on the inside vertical surface of the bowl shell.

## Tubular-Type Purifier

This type of purifier consists essentially of a hollow rotor or bowl, which rotates at high speeds. The rotor has an opening in the bottom to allow the dirty fuel to enter. It also has two sets of openings at the top to allow the fuel and water to discharge. A coupling unit to a spindle connects the bowl, or hollow rotor, of the purifier. The spindle is suspended from a ball bearing assembly. The bowl is belt-driven by an electric motor mounted on the frame of the purifier.

The lower end of the bowl extends into a flexibly mounted guide bushing. The assembly restrains movement of the bottom of the bowl, but it also allows the bowl enough movement to center itself during operation. Inside the bowl is a device consisting of three flat plates that are equally spaced radially. This device is commonly referred to as the *three-wing device*. The three-wing rotates with the bowl and forces the liquid in the bowl to rotate at the same speed as the bowl.

The liquid to be centrifuged is fed, under pressure, into the bottom of the bowl through the feed nozzle.

After the bowl has been primed with water, separation is basically the same as it is in the disc type purifier. Centrifugal force causes clean fuel to assume the innermost position (lowest specific gravity) and the higher density water and dirt are forced outward towards the sides of the bowl. Fuel and water are discharged from separate openings at the top of the bowl. The location of the fuel-water interface within the bowl is determined by the size of a metal ring called a *ring dam*, or by the setting of a discharge screw. The ring dam or discharge screw are also located at the top of the bowl. Any solid contamination separated from the liquid remains inside the bowl all around the inner surface.

Specific instructions for the operation of a purifier should be obtained from the NAVSEA technical manual that is provided with the unit. The following general information applies to the basic operation of purifiers in naval service.

When a purifier is operated as a separator, the bowl must be primed with fresh water before any fuel is admitted to the purifier. The water seals the bowl, and the spinning bowl creates an initial equilibrium of layers of liquid according to specific gravities. If the bowl is not primed, the fuel will be lost through the water discharge ports.

The time required for purification and the output of a purifier depends on many factors. Two important factors are the size of the sediment particles and the temperature of the incoming dirty fuel. In order for any purifier to operate at its rated capacity in gallons per hour, the fuel must be heated to a specified temperature. In this way, the viscosity of the fuel is reduced. (The fuel becomes thinner.) A lower viscosity does two things: (1) it lowers the specific gravity, and (2) it enables the fuel more easily to give up any water, which may be entrained.

The viscosity of the fuel determines largely the length of time required for purification. The more viscous the fuel, the longer the required time is for the fuel to be subjected to centrifugal force. In other words, decreasing the viscosity of the fuel by heating will speed up the purification process and will increase the capacity of the purifier. To reach a higher temperature, the fuel must pass through a heater. In this way, the fuel will reach the proper temperature in the heater before it enters the purifier bowl.

Proper care of any fuel purifier requires that the bowl be cleaned as required and that all sediment be carefully removed. How often you clean a purifier depends on the amount of foreign matter in the fuel to be purified. If the amount of foreign matter in a fuel is not known, you should shut down the machine and check it. The amount of sediment found in the bowl at this time will indicate how often you should clean the purifier. NAVSEA technical manuals, PMS, and the Engineering Operational Sequencing System (EOSS) furnish detailed procedures for operating and maintaining purifiers. Carefully follow these written procedures when you are operating or performing maintenance on purifiers.

It should be obvious from the preceding information that the purpose of the external fuel system is to store and deliver clean fuel to the fuel injection equipment.

**Fuel Injection Systems**

The fuel injection equipment used on Navy diesel engines is the mechanical type. Some engine manufacturers make and install their own fuel injection equipment. Others rely on manufacturers who specialize in fuel injection equipment and who design and modify their products to meet the requirements of the engine manufacturer.

This equipment will vary in construction and method of forcing fuel into the combustion chamber; however, in every case, the fuel injection equipment for any diesel engine must accomplish several basic functions.

## Functions

The five general functions that fuel injection equipment must accomplish are metering the fuel, injecting the fuel, timing the injection process, atomizing the liquid fuel, and creating pressure for dispersion of the fuel. Each of these functions is defined as follows:

1. *METERING*—to measure accurately the amount of fuel to be injected, according to engine speed and load

2. *INJECTING*—to force and distribute the fuel into the combustion chamber

3. *TIMING*—to allow fuel injection into each cylinder to start and stop at the proper time

4. *ATOMIZING*—to break liquid fuel up into tiny particles

5. *CREATING PRESSURE*—to create the high pressure required to force fuel into the pressurized combustion chamber

These functions can easily be recalled by remembering the initials MITAC. All five functions are required for effective combustion of fuel in the cylinders of a diesel engine. Now we will discuss the role of each of these factors and how all of the factors work together.

## Metering

Accurate metering, or measuring, of fuel means that for a given engine speed, setting, and load, the same quantity of fuel must be delivered to each cylinder just before each power stroke of the engine.

If this does not happen, engine speed will be erratic and the horsepower output of the engine will not be uniform. Smooth engine operation and even distribution of the load between cylinders requires that the same amount (volume) of fuel be delivered to a particular cylinder each time it fires, and that equal volumes of fuel are delivered to all cylinders of the engine.

## Injecting

The fuel system on the engine must control the rate of injection, which, in turn, determines the rate of combustion. The rate of fuel injection at the start must be low enough so that excessive fuel does not accumulate in the cylinder during the first phase (physical) of injection delay (before combustion begins). Injection should then proceed at such a rate that the rise in pressure in the combustion chamber is not too great.

However, the rate of fuel injection must be such that the fuel is introduced as rapidly as possible to obtain complete burning of the fuel-air mixture. An incorrect rate of injection affects engine operation in the same way as improper timing. When the rate of injection is too high, the symptoms are similar to those caused when fuel injection is too early. When the fuel injection rate is too low, the symptoms are similar to those caused when fuel injection is too late.

## Timing

In addition to measuring the amount of fuel and rate of fuel injection, the fuel injection equipment must cause these events to occur at the proper time. Correct timing is vital to ensure that complete combustion takes place and that maximum energy is obtained from the fuel. (In other words, the engine develops rated horsepower for each pound of fuel burned.) When fuel is injected too early in the cycle, ignition may be delayed because the temperature of the air charge in the cylinder is not high enough. On the other hand, late injection results in rough, noisy operation of the engine.

Noisy engine operation occurs when the engine cannot convert as much energy from the fuel into the horsepower required to move the load. Late injection permits some fuel to be wasted by wetting of the cylinder walls and piston crown. This condition, of course, results in poor fuel economy, higher than normal exhaust gas temperatures, and smoky exhaust.

## Atomizing

Shortly before the top of the compression stroke, at a point controlled by the mechanical injection timing arrangement, one or more jets of fuel are introduced into the combustion chamber. As explained previously, ignition of fuel does NOT occur immediately on injection. The fuel droplets absorb heat from the compressed air swirling around the combustion chamber. This process is necessary because it causes the liquid fuel to *vaporize* so it can burn! The duration of the second phase (chemical) of ignition delay is controlled by the design (shape) of the combustion chamber, fuel and air inlet temperatures, degree of atomization of the fuel, and the quality of the fuel. When the fuel-air mixture reaches a temperature at which self-ignition occurs, the flame begins to spread. Injection of the remaining volume of fuel for the cylinder continues during this time. The ignition delay period must be short so that diesel knock can be avoided.

Once the flame has been completely initiated, the fuel being delivered to the cylinder is that which is being injected into the burning mixture. This fuel vaporizes and burns almost instantaneously. This process is the third phase of fuel injection. Liquid fuel must be injected into each cylinder in the form of a fine spray. Proper atomization increases the surface area of the fuel, which must be exposed to oxygen molecules in the air so that complete burning of the fuel can take place and the rated horsepower can be developed. To avoid simultaneous combustion of all droplets of the fuel spray (detonation), the injected spray is usually in the form of fine droplets to start ignition (beginning of second phase) and larger droplets later in the phase. The degree of atomization of the fuel is controlled by the diameter and shape of the nozzle orifice(s) or opening(s), injection pressure, and the density of the air charge in the combustion chamber.

## Creating Pressure

The quality (volume) of fuel, the rate at which it is injected into the cylinders of an engine, and the timing and duration of the injection event are all controlled by the fuel injection equipment. At the beginning of injection, fuel pressure may be as low as 1,800 psi to as high as 30,000 psi, depending upon the design of the equipment. The fuel injection equipment must raise the pressure of the fuel enough to overcome the force of the compressed air charge in the combustion chamber and ensure proper dispersion (distribution) of the fuel being injected into the combustion space. Proper dispersion of the atomized fuel in the air charge is an important factor for complete combustion to take place. The atomization process and the penetration of the fuel, which determines the distance through which the fuel droplets travel after leaving the injector tip or nozzle, affect Dispersion of the fuel, in part. If the atomizing process results in fuel droplets that are too small, they will not have sufficient weight to penetrate very far into the air charge. Too little penetration results in the fuel igniting and burning before it is properly dispersed through the air charge in the combustion space. Since penetration and atomization tend to oppose each other, a compromise in the degree of each is necessary in the design of the fuel injection equipment.

## Fuel Injection Equipment

There are three methods commonly used for the mechanical injection of fuel (at the proper amount, time, and duration) into the cylinders of a diesel engine. These methods are as follows:

1. Pump controlled (jerk pump)

2. Distributor

3. Unit injector

NOTE: A fourth method, known as pressure time (PT) uses unit injectors. This method is unique to Cummins diesel engines and is not considered to be common; therefore, it will not be explained in this rate training manual. The three methods listed above will be explained in the sections that follow.

**Jerk Pump Fuel Injection System**

Jerk pump fuel injection systems consist of high-pressure pumps and pressure-operated spray valves or nozzles that are separate components. In some engines, such as the Alco, there is only one pump and one nozzle for each cylinder. In other engines, such as the Fairbanks-Morse opposed piston engine, each cylinder has two pumps and two nozzles. The pump itself carries out most of the injection event. The pump raises pressure, meters the fuel, and times the injection. The nozzle is simply a spring-loaded check valve that reacts to the pressure supplied from the high-pressure pump.

NOTE: A major manufacturer of jerk pump fuel injection systems is the American Bosch company. The system may use either of two different types of pumps, designated APF or APE. The letter *F* in APF identifies a pump that does not have its own drive, and the letter *E* in APE indicates a pump with a self-contained drive.

**Bosch APF Pump**

Type APF pumps are of the single-cylinder design with the plunger pump for each cylinder contained in separate housing. In a 6-cylinder engine, there are six separate APF pumps. Each pump is cam driven and the setting of the control rack regulates fuel volume.

## Bosch APE Pump

Type APE pumps are assembled with all the individual cylinder plungers in a single housing. The injection pumps operate from a single camshaft in the bottom part of the housing. The cam lobes are arranged so that the firing order is consistent with the engine firing order. Each revolution of the camshaft provides one fuel charge from each outlet.

Although our discussion will be on the APE pump system, the APF pump operates on the same principles. The information on the pumping principle, metering principle, and delivery valve operation also applies to the APF pump.

## General Operating Principles

The supply pump is a plunger-type, cam activated pump that is equipped with inlet and outlet check valves and a hand-priming pump. The supply pump draws fuel from the fuel tank through a primary filter and discharges the fuel to a final stage filter. An arrow stamped on the supply pump housing to indicate the direction of fuel flow. From the final stage filter, the fuel then flows into the injection pump sump (fuel gallery) which provides all six plunger and barrel assemblies with fuel. An overflow valve, which regulates and maintains the pump sump (gallery) pressure, is attached to the injection pump housing. The overflow valve assembly also permits any air in the system to work its way out and return to the fuel tank.

When the pump plunger is in its lowest position, fuel from the sump (gallery) enters the barrel ports and fills the volume above the plunger. The upward movement of the plunger (through cam action) seals off the barrel ports and forces fuel (now under high pressure) through the delivery valve assembly and high pressure tubing to the holder and nozzle assembly.

Fuel entering the nozzle holder inlet flows through passages to the nozzle, which contains a spring-loaded valve. Fuel pressure exerts a force against the lower end of the valve, which is opposed by the spring force. At a preset pressure, the spring force is overcome and the valve rises, as a result permitting the fuel to flow through the nozzle spray holes (orifices) and into the combustion chamber.

The slight fuel leakage between the nozzle valve and body (required for nozzle lubrication) is returned to the fuel tank through a leak-off line. Normally, the nozzle leak-off line is connected to the pump return line; consequently, both are returned to the fuel tank in a single line.

So far, our discussion has been general. We will now discuss in more detail the principles by which the fuel is pumped, metered, and delivered to the nozzle.

## Pumping Principle

We will first discuss the pumping principle behind the action of the APE fuel pump as it pump fuel. When the plunger is at the bottom of its stroke, fuel from the pump sump flows through the barrel ports and fills the volume above the plunger. The sump fuel initially fills the vertical slots and connecting cutaway areas of the plunger.

Upward movement of the plunger seals off the barrel ports, thus trapping fuel in the barrel. Additional upward movement of the plunger forces fuel through the delivery valve, high-pressure tubing, nozzle, and finally to the combustion chamber. Fuel delivery will stop when the plunger helix uncovers the barrel port. This action releases the trapped fuel through the slot in the plunger and out through the barrel ports.

**Metering Principle**

The positioning of the plunger helix is automatically controlled by movement of the speed governor output shaft, which is attached to the rear of the pump housing. The governor, through linkages, positions the control racks that rotate the segment gears and control sleeves, which, in turn, radially position the plunger flange and helix. The fuel control rod (rack) teeth engage the plunger gear teeth to control fuel metering. Lateral movement of the fuel rack causes the plungers to rotate. This action, in turn, determines the effective stroke of the plunger-by-plunger helix position in relation to the barrel port.

The amount of fuel delivered is controlled by the position of the *plunger helix*. When the plunger is rotated to a certain position, the effective part of the stroke (that position of the stroke from the closing of the barrel ports by the top of the plunger to the point where the edge of the helix raises above the barrel ports) is long, and fuel delivery is at a maximum. When the plunger is rotated to a different position, the effective part of the stroke is reduced since the helix will uncover the barrel ports sooner (at a lower position). This action reduces the volume of fuel delivered to the cylinder. When the vertical slot on the plunger is in line with one of the barrel ports, there is no effective stroke and therefore no fuel delivery.

**Delivery Valve Operation**

The delivery valve assembly is located directly above the plunger. The delivery valve assembly assists the injection function by preventing irregular losses of fuel from the delivery to the supply side of the system between pumping strokes.

The delivery valve assembly consists of a valve with a conical seat and a valve body with a corresponding mating seat. Opening pressure is controlled by the force of the delivery valve spring that engages the top of the valve.

Since liquid fuel trapped in the barrel is essentially incompressible, pressure is created after the plunger, on its upward stroke, closes off the barrel ports. When this hydraulic pressure overcomes the force of the delivery valve spring, the valve opens and fuel passes through it into the injection tubing.

When the edge of the plunger helix passes the lower edge of the barrel port, there is a rapid drop in fuel pressure below the delivery valve, and the force of the valve spring, combined with the high differential in pressure, begins to return the valve to its seat.

As the valve starts downward into the valve body, the lower edge of its retraction piston enters the valve bore. At that moment, the flow of fuel by the delivery valve stops. The continued downward movement of the valve (retraction piston) increases the volume on the high-pressure side by the amount of the piston retraction (its displacement volume) and, consequently, reduces the residual pressure in the injection tubing and nozzle holder. This lower pressure promotes rapid closing of the injection nozzle valve and diminishes the effect of hydraulic pressure waves that exist in the tubing between injections, thereby minimizing the possibility of the nozzle reopening (a secondary injection) before the next regular delivery cycle.

## Plungers

Plungers include three different types of helixes-a lower helix, an upper helix, and both an upper and a lower helix. (The shape and the position of the helix governs the fuel delivery curve, the beginning or ending of injection, or both the beginning and ending of injection.)

For the lower helix plunger the effective stroke (injection) always begins at the same time, regardless of where the plunger is rotated because the top end that closes the ports is flat. The end of injection can be varied because of the sloping design of the helix. Injection has a constant beginning and a variable ending. This type of plunger is used in pumps marked "timed for port closing."

For the upper helix plunger, the beginning of the effective stroke varies as the plunger is rotated because the top edge, which closes the ports, is sloping. Thus, the beginning of delivery is variable and the ending is constant. Plungers of this type are used in pumps marked "timed for port opening."

A third type of plunger has a variable beginning (upper helix) and a variable ending (lower helix) design. Rotation of this type of plunger varies both the beginning and the ending of delivery of the fuel.

**Distributor Fuel Injection System**

In our discussion of the distributor fuel injection system, we will use the DPA type for our example because of its wide use on Navy small craft. The DPA pump is a compact unit that is lubricated throughout by fuel and requires no separate lubrication system. It contains no ball or roller bearings, gears, or highly stressed springs. A governor that is either mechanically or hydraulically operated within the pump itself maintains sensitive speed control.

**Design and Components**

In the DPA distributor type of injection pump, the fuel is pumped by a single element. A rotary distributor that is integral with the pump distributes the fuel charges in the correct firing order. Therefore, equality of delivery to each nozzle is an inherent feature of the pump. Since the timing interval between injection strokes is determined by the accurate spacing of the distribution ports and the operating cams do not have to be adjusted, accurate timing of delivery is also an inherent feature of the pump.

There is a central rotating steel member known as the pumping and distributing rotor. The rotor is a close fit in a stationary steel cylindrical body, called the hydraulic head. The pumping section of the rotor has a transverse bore containing two opposed pump plungers. These components rotate inside a cam ring in the pump housing, and operate through rollers and shoes sliding in the rotor. The cam ring has as many internal lobes as the engine has cylinders. The opposed plungers have no springs but are moved outward hydraulically by fuel pressure.

The pumping and distributing rotor is driven by splines from a drive shaft. At its outer end, the rotor carries a vane-type fuel transfer pump. With a piston-type regulating valve housed in the end plate, the transfer pump serves to raise the pressure of the fuel to an intermediate level, known as the transfer pressure.

## Operation

The DPA pump is driven at half engine speed. As the rotor turns, a charging port in the rotor aligns with the metering port in the hydraulic head. Fuel at metered pressure then flows into the central passage in t the rotor and forces the plungers apart. The amount of plunger displacement is determined by the amount of fuel that can flow into the element while the ports are aligned. The fuel inlet port closes as rotation continues. As the rotor turns, the fuel remains isolated into the rotor. As the single distributor port in the rotor comes into alignment with one of the outlet ports in the hydraulic head, the plungers are forced quickly together by the action of the cam. At this point, high pressure is generated, and the pressurized fuel passes via a high-pressure line to an injector. From the injector, the fuel passes to the engine combustion chamber. This entire cycle of operation is repeated once for each engine cylinder per pump revolution.

## Injection Nozzles

An injection nozzle assembly serves to position the nozzle accurately in the engine cylinder. An injection nozzle assembly will contain the necessary spring and pressure adjustment means to provide for the proper action of the nozzle valve. An injection nozzle assembly will also provide a means by which fuel can be conducted the nozzle and the combustion chamber of the engine. Although manufacturers produce a wide variety of nozzles to meet the requirements of several different combustion systems and engine sizes, there are essentially two basic groups of injection nozzles: the pintle nozzles and the hole nozzles.

The nozzle holder and the pintle-type nozzle is responsible for high-pressure fuel delivery. The high-pressure fuel from the injection pump enters the nozzle holder body through a metal-edge strainer. From the strainer, the fuel goes through a drilled fuel passage that extends to the bottom of the nozzle holder body. The nozzle, with its spray tip, is held against the bottom of the nozzle holder by the cap nut. A groove in the top of the nozzle forms a circular passage for the fuel between the nozzle and the holder.

Several vertical ducts carry the fuel from the circular passage to the fuel cavity, near the bottom of the nozzle. The nozzle valve cuts in sharply to a narrower diameter in the fuel cavity, providing a surface against which the high-pressure fuel in the fuel cavity can act to raise the valve from its seat in the spray tip. When the valve is raised from its seat, the fuel sprays out to the combustion chamber through a ring of small holes (hole nozzle), or around the pintle and out through the single hole (pintle nozzle).

The valve has a narrow stem that projects into the central bore of the nozzle holder where it bears against the bottom of the spindle. The spindle is held down by the pressure-adjusting spring. Whenever the upward force of the high-pressure fuel acting on the needle valve exceeds the downward force of the spring, the valve can rise. The moment the spring force is greater, the valve will snap back to its seat. A pressure-adjusting screw or shims regulate the spring tension.

Regardless of the close-lapped fit of the valve, some fuel will leak past the valve and rise through the central bore of the nozzle holder. This fuel lubricates the moving parts and carries away heat from the injector. The bypassed fuel then drams off through the fuel dram connection to a drip tank. The bleeder screw can serve to bypass fuel to the nozzle, sending the fuel directly to the fuel drain.

**Pintle Nozzle**

The valve of the pintle nozzle has an extension that protrudes through the hole in the bottom of the nozzle body and produces a hollow cone-shaped spray. The included angle of the spray cone may be up to a maximum of 60 degrees, depending on the type of combustion chamber in which it is used. A pintle nozzle generally opens at a lower pressure than the pressure at which the hole nozzle opens because fuel flows more readily from the large hole of the pintle nozzle. Although atomization of the fuel is not so complete in the pintle nozzle as it is in the hole nozzle, penetration into the combustion space is greater. Consequently, pintle-type nozzles are used in engines having pre-combustion, divided, air cell or energy-cell combustion chambers, where mixing of fuel and air is largely dependent on combustion reaction or turbulence. The motion of the pintle tends to inhibit the formation of carbon crust on the tip of the nozzle.

**Multiple-Hole Nozzle**

The multiple-hole nozzle provides good atomization but less penetration than the pintle nozzle. The multiple-hole nozzle is used with the open type of combustion chamber in which high atomization is more important than penetration. The spray pattern of the hole nozzle is dependent on the number and placement of the holes or orifices.

Spray openings, or orifices, are from 0.006 inch up to about 0.033 inch in diameter, and their number may vary from three to as many as 18 for large bore engines.

Regardless of design, all nozzles and tips function to direct the fuel into the cylinder in a pattern that will bring about the most efficient combustion. Obviously, the slightest defect in nozzles and tips will have an adverse effect on engine operation.

**Unit Fuel Injector System**

Unlike the design of the other two fuel injection systems, the unit injector provides each cylinder with its own high-pressure pump. The unit injector design eliminates the need for a remotely located high-pressure pump and high-pressure external fuel lines, such as those used in the systems we have covered in the preceding sections. The unit fuel injector (UFI) is complete with pumping and timing element, fuel control, and injection valve spray tip assembly to control the quantity, rate, and timing of fuel delivery. The UFI is used on all General Motors engines; therefore, there are various models available to meet the needs of the engines. Our discussion will cover the two most commonly used types: (1) the crown valve and (2) the needle valve injector.

The crown valve injector was placed in service in 1953 and is still in use within the Navy. In 1962, the needle valve injector incorporating a new tip design was introduced. The needle valve injector provides for improved economy and emissions by increased pop pressure (2300 to 3300 psi for the needle valve injector compared with 450 to 850 psi for the crown valve injector) and a more precise method of fuel control than the crown valve type of unit injector.

## Design and Components

The unit injector is installed in the cylinder head. It is held in place by an injector clamp. The cylinder head has a copper tube into which the injector fits snugly with the spray tip projecting slightly into the cylinder clearance space. Water circulates around the copper tube and cools the lower part of the injector. Two fuel lines are connected to each injector; one carries fuel to the injector and the other carries away the fuel that is bypassed. The injector is operated by a rocker arm and push rod assembly, which work off the camshaft. The amount of fuel injected is regulated by the control rack, which is operated by a lever secured to the control tube.

## Operation

Fuel, under pressure, enters the injector at the inlet side through a filter cap and filter. From the filter, the fuel passes through a drilled passage into the supply chamber, the area between the plunger bushing and the spill deflector, in addition to that area under the injector plunger within the bushing. The plunger operates up and down in the bushing, the bore of which is open to the fuel supply in the annular ring-shaped chamber by two funnel-shaped ports in the plunger bushing.

The motion caused by the injector rocker arm is transmitted to the plunger by the follower, which bears against the follower spring. In addition to the reciprocating up-and down motion, the plunger can also rotate around its own axis by the gear, which meshes with the control rack. For the metering of the fuel, an upper helix and a lower helix are machined in the lower part of the plunger. The relation of the helices to the two fuel ports will change as the rotation of the plunger changes.

As the plunger moves downward (due to the force of the injector rocker arm), a portion of the fuel trapped under the plunger is displaced into the supply chamber through the lower port. This action will occur until the port is closed off by the lower end of the plunger. A portion of the fuel trapped below the plunger is then forced up through a central passage in the plunger into the fuel metering recess and into the supply chamber through the upper port until that port is closed off by the upper helix of the plunger. With the upper and lower ports both closed off, the remaining fuel under the plunger is subjected to increased pressure by the continuing downward movement of the plunger.

When sufficient pressure builds up, the flat, non-return check valve opens. The fuel in the check valve cage, spring cage, tip passages, and tip fuel cavity is compressed until the force of the pressure acting upward on the needle valve is sufficient to open the valve against the downward force of the valve spring. As soon as the needle valve lifts off its seat, the atomized fuel is forced through the small orifices in the spray tip and atomized into the combustion chamber. When the lower land of the plunger uncovers the lower port in the bushing, the fuel pressure below the plunger is relieved. The valve spring then closes the needle valve (or injector valve), and injection stops. A pressure relief passage is provided in the spring cage. This passage permits any bleed-off of fuel that may leak past the needle pilot in the tip assembly. The check valve, located directly below the bushing or mounted in the spray tip, prevents leakage from the combustion chamber into the fuel injector in case the valve is accidently held open by a small particle of dirt. The injector plunger is then returned to its original position by the injector follower spring.

On the return upward movement of the plunger, the high-pressure cylinder within the bushing is again filled with fuel through the ports. The constant circulation of fresh, cooler fuel through the injector renews the fuel supply in the chamber, helps carry heat from the injector, and also effectively removes all traces of air that might otherwise accumulate in the system and interfere with accurate metering of the fuel. The fuel injector outlet opening, through which the excess fuel returns to the fuel return manifold and then back to the fuel tank, is directly adjacent to the inlet opening.

A change in the position of the helices, by rotation of the plunger, retards or advances the closing of the ports and the beginning and ending of the injection period. At the same time, a change in the position of the helices increases or decreases the amount of fuel injected into the cylinder. With the control rack pulled out all the way, the upper port is not closed by the helix until after the lower port is uncovered. Consequently, with the rack in this position, all of the fuel is forced back into the supply chamber and no injection of fuel takes place. With the control rack pushed all the way in (full injection), the upper port is closed shortly after the lower port has been covered, thus a maximum effective stroke and maximum injection is produced. From this no injection position to the full injection position (full rack movement), the contour of the upper helix advances the closing of the ports and the beginning of injection.

The injection system must contain pure fuel. The unit injectors have spray tip holes as small as 0.005 inch. Pressures of approximately 20,000 psi are developed by a combination of spray hole area restriction and plunger/bushing design. Typically, the plunger and bushing are matched to a diametrical clearance of 60 millionths of an inch to prevent leakage during the injection cycle. As you can see, impurities of any kind can damage the unit.

**Purging the Diesel Engine Fuel Injection System**

When an engine fails to operate, stalls, misfires, or knocks, there may be air in the high-pressure pumps and in the fuel lines. Unlike liquid fuel, which is incompressible, when air is present in the system, compression and expansion of air will occur and the injector valves will either fail to open or will not open at the proper time. Diesel engineers can determine the presence of air in a fuel system by bleeding a small amount of fuel from the top of the fuel filter or by slightly loosening an air bleeder screw or plug. If the fuel appears quite cloudy, it is likely that there are small bubbles of air in the fuel. When working with fuel systems, they remember that if air is entering a fuel line, the pressure within the fuel line must be lower than atmospheric pressure. The smallest of holes in the transfer pump suction piping will allow enough air to flow into the system to air bind the high-pressure pumps. Carefully inspect all fittings in the suction piping. A loose fitting or a damaged thread condition will allow air to enter the system. On installations where flanged connections are used, be sure to check the condition of the gaskets. Inspect tubing (especially copper) and flexible hose assemblies carefully for cracks that may result from constant vibration or rubbing.

The use of tubing and flexible hose assemblies on diesel engine fuel systems is common. Diesel engineers will find that flexible hose assemblies are used more on the supply or low-pressure side of the injection equipment while tubing is more commonly used on the high-pressure side of the injection equipment. The use of tubing and flexible hose assemblies is also the means by which all pressure gauges of a diesel may be located on a central gauge board away from the system the gauges are monitoring.

If an engine is allowed to run out of fuel, you can expect trouble from air that enters the fuel system. If there is a considerable amount of air in the filter, a quick method of purging the system of air is to remove the filling plugs on top of the filter and pour in clean fuel until all air is displaced. Air remaining in the system is then removed by using the hand-priming pump. Open the system between the pump and the filter. Operate the hand-priming pump until all air is removed and only clear fuel flows from the line. Then close the line. Repeat the same procedure at other points in the system, such as between strainers and the filters, between the filters and the high-pressure pumps, and at the overflow line connection (excess fuel return line) on the high-pressure pump housing. In small, high-speed diesel engines, you may need to prime only at the overflow connection. Since priming high-pressure lines is time-consuming, attempt to start the engine before purging these lines. However, do not crank the engine for more than the specified interval of time. If the engine still fails to start, the high-pressure lines should be primed. Since the procedure necessary to prime high-pressure lines will vary considerably with different installations, follow the NAVSEA technical manual instructions for the proper procedure.

## Control Devices in a Diesel Engine

In this section of the chapter, we will discuss the methods and the devices that serve to control the output of the injection pumps and injectors. By controlling the output of the fuel injector system, these devices ensure control of engine operation.

## Governors

A governor is a speed-sensing device on an engine that serves to maintain a constant speed in revolutions per minute (rpm) within the design power rating of a diesel engine. A governor may also serve to limit high and low-idle revolutions per minute of the engine. All governors used on diesel engines control engine speed by regulating the amount of fuel delivered to the cylinders. An example of engine control without the use of a governor is in the conventional automobile—the driver senses the engine speed and engine load changes. The driver's movement of the throttle is an extension of a conditioned reflex; in other words, the driver acts as the governor. However, the driver will not be capable of reacting to load and speed changes quickly enough insofar as the diesel engine is concerned. This is because diesel engines can maintain a constant speed better using a governor. In diesel engines, the speed and power output is determined by varying the amount of fuel injected into the cylinders to control combustion. The governor acts through the fuel injection equipment to regulate the amount of fuel delivered to the cylinders. As a result, the governor holds engine speed reasonably constant during fluctuations in load. Hydraulic and mechanical are the two principal types of governors. To understand why different engines use different governors you will need to know the meaning of several terms.

## Terminology

Before considering the operating principles of various types of governors, you should become familiar with the following terms:

1. *SPEED DROOP* is the decrease in speed of the engine from a no-load condition to a full load condition. Speed droop is expressed in rpm or (more commonly) as a percentage of normal or average speed.

2. *ISOCHRONOUS GOVERNING* is maintaining the speed of the engine truly constant, regardless of the load. This means governing with perfect speed regulation or zero speed droop.

3. *HUNTING* is the continuous fluctuation (slowing down and speeding up) of the engine speed from the desired speed. Governor causes hunting from over control.

4. *STABILITY* is the ability of the governor to maintain the desired engine speed without fluctuations or hunting.

5. *SENSITIVITY* is the change in speed required before the governor will make a corrective movement of the fuel control mechanism and is expressed as a percentage of the normal or average speed.

6. *PROMPTNESS* is the speed of action of the governor. It identifies the time interval required for the governor to move the fuel control mechanism from a no-load position to a full-load position. Promptness depends on the power of the governor; the greater the power, the shorter the time required to overcome the resistance.

7. *Surges* is rhythmic variations of speed of large magnitude. They can be eliminated by blocking the fuel linkage manually and will not reappear when returned to governor control unless the speed adjustment of the load changes.

8. *Jiggles* are high-frequency vibrations of the governor fuel rod end or engine fuel linkage. Do not confuse jiggle with the normal regulating action of the governor.

Now that you have read the definitions of some of the terms associated with governors, we will proceed with our comparison between mechanical and hydraulic governors.

## Classification of Governors

Diesel engines used by the Navy have two general classes of governors— *mechanical* and *hydraulic*. Both types of governors serve to regulate engine speed by controlling the fuel injected into the engine is referred to as *speed-regulating governors*. The mechanical governor is usually simple in design, contains few parts, and is relatively inexpensive. It is frequently used as a speed-limiting device or when extremely sensitive operation is not required.

The hydraulic governor is more complex in design and contains more parts than the mechanical governor contains. However, it is more sensitive to speed variations, quicker acting, and more accurate because of the reduction of friction and the small mass of moving parts. Due to the forces exerted by the hydraulic system, a hydraulic governor can be smaller and more compact than a mechanical governor is and still operate the fuel mechanism of large engines.

Both mechanical and hydraulic governors are described in detail later in this chapter. Governors may also be classified according to the function or functions they perform, the forces they use in operation, and the means by which they operate the fuel control mechanism. The load on the engine determines the function of a governor on a given engine and the degree of control required. Governors are classified according to their function as constant speed, variable-speed, speed limiting, and load limiting.

Some installations require a constant engine speed from a no-load condition to a full-load condition. Governors that maintain one speed, regardless of load, are called *constant speed* governors. Governors that maintain any desired engine speed over a wide speed range and that can be set to maintain a desired speed in that range are classified as *variable-speed* governors. Speed-control devices that serve to keep an engine from exceeding a specified maximum speed and from dropping below a specified minimum speed are classified as *speed limiting* governors. Some speed-limiting governors limit maximum speed only. Some engine installations need a control device to limit the load that the engine will handle at various speeds. Such devices are called *load limiting* governors. Some governors are designed to perform two or more of these functions.

## Mechanical Governors

A mechanical governor controls the speed of the engine by controlling the spring-balanced position of the flyweights. When the load is decreased or removed from the engine (such as when a clutch is disengaged) and the speed exceeds its former balanced setting, the increased speed of the flyweights develops a greater centrifugal force that upsets the former flyweight spring balance. A new balance is achieved by the weights as they move outward and further compress the spring. Any movement of the flyweights is reflected in a vertical change in position of link A. When the load is increased on the engine, the fuel is injected will be inadequate for the increased load and the engine will slow down. The centrifugal force of the flyweights will then decrease and permit the former balanced spring force to move a link down until the new flyweight position again is balanced by the spring. You should note that the linkage movement causes an increase in fuel when the load is increased and a decreased supply of fuel as the load is reduced. From this discussion, it is evident that the mechanical governor controls the fuel supply by virtue of the flyweight position.

**Hydraulic Governors**

Hydraulic governors are speed-sensitive elements. The hydraulic governor depends on a flyweight arrangement similar to the mechanical governor. However, the power supply that moves the fuel mechanism is operated hydraulically rather than through direct mechanical linkage with the flyweights. The flyweights of the hydraulic governor are linked directly to a small pilot valve that opens and closes ported passages, admitting oil under pressure to either side of a power piston that is linked to the fuel control mechanism. Since the flyweights move only a lightweight pilot valve, the inherent design of the hydraulic governor is more sensitive to small speed changes than the design of the mechanical type of governor, which derives all of its working power from the flyweights. The larger and heavier the fuel control mechanism, the more important it is to employ a hydraulic governor.

There is always a lag between a change in fuel setting and the time the engine reaches the new desired speed. Even when the fuel controls are set as required during a speed change, hunting caused by overshooting will occur. As long as engine speed is above or below the desired new speed, the simple hydraulic governor will continuously adjust (overcorrect) the fuel setting to decrease or increase the delivery of fuel. For this reason, a hydraulic governor must have a mechanism that will discontinue changing the fuel control setting slightly before the new setting has actually been reached. This mechanism, used in all modern hydraulic governors, is called a *compensating device*.

The buffer piston, buffer springs, and needle valve in the hydraulic circuit between the control land of the pilot valve plunger and the power piston comprise the buffer compensating system of the governor. Lowering the pilot valve plunger permits a flow of pressurized oil into the buffer cylinder and power cylinder. This flow of oil moves the power piston up to increase fuel. As the pilot valve plunger moves up, oil is permitted to flow from the buffer

41

cylinder and power cylinder to the governor sump, and the power piston spring moves the power piston down to decrease fuel. The rate of compensation is adjusted by regulating the oil leakage through the compensating needle valve. If the compensating needle valve is adjusted correctly, only a slight amount of hunting will occur after a load change. This hunting will quickly be dampened out, resulting in stable operation through the operating range of the governor.

## Overspeed Safety Devices

Engines that are maintained in proper operating condition seldom reach speeds above those for which they are desired. However, conditions may occur to cause excessively high operating speeds, such as when a ship's propeller comes out of the water in rough seas. Operation of a diesel engine at excessive speeds is extremely dangerous because of the relatively heavy construction of the engine's rotating parts. A high-speeding engine develops inertial and centrifugal forces that may seriously damage parts or even cause them to fly apart. Therefore, *why* an engine may reach a dangerously high speed must be known and *how* to bring it under control when excessive speed occurs.

In some two-stroke cycle engines, lubricating oil may leak into the cylinders because of leaky blower seals or broken piping. Even though the fuel is shut off, the engine may continue to operate, or even "run away," because of the combustible material coming from the uncontrolled source. Engines in which lubricating oil may accumulate in the cylinders generally have an automatic mechanism that shuts off the intake air at the inlet passage to the blower. If there is no air shutoff mechanism and if shutting off the fuel will not stop an engine that is overspeeding, a cloth article such as a blanket or a pair of dungarees should be placed over the engine's intake to stop airflow. This action will subsequently stop the engine.

Excessive engine speeds are more commonly found where there is an improperly functioning regulating governor than where lubricating oil accumulates in the cylinders. To stop an engine that is overspeeding because of lubricating oil in the cylinders, stop the flow of intake air. To accomplish an emergency shutdown or reduction of engine speed when the regulating governor fails to function properly, shut off or decrease the fuel supply to the cylinders. The fuel supply to the cylinders of an engine can be shut off in several ways, either manually or automatically:

1. Force the fuel control mechanism to the NO FUEL position.

2. Block the fuel line by closing a valve.

3. Prevent the mechanical movement of the injection pump.

Overspeed safety devices automatically operate the fuel and air control mechanisms. As emergency controls, these safety devices operate only in case the regular speed governor fails to maintain engine speed within the maximum design limit. Devices that bring an overspeeding engine to a full stop by completely shutting off the fuel or air supply are called *overspeed trips*. Devices that reduce the excessive speed of an engine, but allow the engine to operate at safe speeds, are more commonly called *overspeed governors*.

**Overspeed Governors and Trips**

All overspeed governors and trips operate on a spring-loaded centrifugal governor element. In overspeed devices, the spring tension is great enough to overbalance the centrifugal force of the weights until the engine speed rises above the desired maximum. When the speed setting of the governor is reached, the centrifugal force overcomes the spring tension and operates the mechanism that stops or limits the fuel or air supply.

When a governor serves as a safety device, the fuel or air control mechanism is operated by the centrifugal force either directly, as in a mechanical governor, or indirectly, as in a hydraulic governor. In an overspeed trip, the shutoff control is operated by a power spring. The spring is placed under tension when the trip is manually set and is held in place by a latch. If the maximum speed limit is exceeded, a spring-loaded centrifugal weight will move out and trip the latch, allowing the power spring to operate the shutoff mechanism.

NOTE: If the engine overspeeds and exceeds its rated rpm trip setting, internal inspection of the engine must be accomplished before the engine is restarted.

Overspeed safety devices must always be operative and must never be disconnected for any reason while the engine is operating. All overspeed safety devices should be tested under the Planned Maintenance System (PMS).

**Basic Care of the Governor**

Contaminants and foreign matter in the governor oil are the greatest single source of governor troubles. Use only new or filtered oil. Be sure that all containers used for the governor oil are clean. The time interval between governor oil changes depends upon many factors: type of service, operating temperature, quality of oil, and so forth. Anytime the governor oil appears to be dirty or breaking down from contaminants or excessive temperatures, drain the governor while it is hot, flush it with the lightest grade of the same oil, and refill it with fresh oil. In any event, follow the PMS for regular oil drain intervals.

A governor should operate several years before needing replacement if it is kept clean and if the drive from the engine is smooth and free from torsional oscillations. Except for isolated cases, so rare they can be almost disregarded, governors do not suddenly fail or break down. Instead, they wear gradually, and give an external indication of their condition in the form of slight hunting, sluggish operation, and so forth. Further deterioration is at a slow enough rate so that an exchange governor may be ordered for installation and the governor can be changed out at a convenient downtime.

**Propulsion Control Systems**

In modern propulsion systems, an integrated system of pneumatic, hydraulic, and electric circuits provides control of the speed and direction of the propeller shaft. Each control system function may use only one of these three mediums, a combination of two, or a combination of all three. The choice of control medium in each instance is based on the performance of a given control function in the most reliable and efficient manner. In general, the different control mediums are used in the following functions:

1.  The basic control medium is pneumatic, and pneumatic circuits perform the majority of propulsion control functions.

2.  Hydraulic circuits are used whenever a large amount of control element actuating power is required, such as when pitch is applied to controllable pitch propellers.

3.  Electric circuits are used extensively for the sensing and indicating of control system conditions and for alarm systems.

The basic requirements for a propulsion control system are threefold. First, it must control the main engines' load, keeping the engines equally loaded. Next, it must maintain propeller shaft speed and direction. Finally, it must maintain the desired pitch since gas turbine ships and most diesel-driven ships have controllable pitch propellers.

The engine speed and load are controlled by the governor of each main engine, through a pneumatic signal sent to the governor, which increases or decreases the tension of the speeder spring.

Control of the propeller shaft speed is done by control of the main engine speed as stated previously. Clutches and reduction gears control propeller shaft direction of rotation, ahead or astern, on non-controllable pitch propeller ships.

In ships with controllable pitch propellers, the pitch is controlled by a signal, either pneumatic or hydraulic, that is sent to an oil distribution (OD) box. There, the signal is converted to a high-pressure hydraulic force, which actuates the propeller blades through a piston and cylinder assembly in the propeller hub.

In most ships with propulsion control systems, the machinery can be operated from three different locations. Local control is usually from a panel mounted on or near the machinery to be operated. The local control station is used for the operation of a single unit, such as one main engine, or for setting the pitch on one propeller. The enclosed operating station (EOS) has a console for the operation of one complete propeller shaft, including main engines, propeller pitch control unit, clutches, and all other machinery required for propulsion. On large ships, there may be one EOS for each propeller shaft. On smaller ships, one EOS is used to control and monitor all propulsion machinery for the ship. The third operating station is the pilothouse console, which controls propeller shaft speed and pitch or direction.

Generally, this station cannot control the starting or stopping of main engines, operate clutches, or control other individual pieces of propulsion machinery. Both the EOS console and pilot-house console will have instruments that indicate shaft rpm, propeller pitch, and other indicators required for the monitoring of the propulsion plant.

Normally, propulsion control systems should operate trouble-free for many years with a minimum of care.

Pneumatic systems need a constant source of clean dry air to operate correctly. If the supply of air is dirty or contains oil or water, the various control valves throughout the system will stick and cause malfunctions. All connections should be checked periodically. If leaks should start, a drop in line pressure will create a faulty signal. Any leaks in a pneumatic system can be located by brushing a solution of soapy water on the connectors of the system. Bubbles that will form at the site of the leak will indicate an air leak.

The worst enemy of any hydraulic system is dirt. Dirt that is allowed to enter the system, either when oil is being added or when other work is being performed, will create problems. Dirt will cause the extremely close clearances of parts in hydraulic components to become damaged. Also, dirt will cause the valves in the system to malfunction.

Electrical systems are all but trouble-free with little routine maintenance required. Occasionally, problems occur because of loose connections or from component failure.

Most propulsion control systems will use a combination of all three mediums: pneumatic, hydraulic, and electric. In troubleshooting a malfunction, only one medium at a time should be checked until the trouble can be isolated and the necessary repairs can be made.

## SUMMARY

Diesel engineers are familiar with factors related to combustion and how these factors affect diesel engines. They know the meaning of turbulence and precombustion and the significance of each to the combustion process of a diesel engine.

In our discussion of fuel systems, we can never overstress the importance of clean fuel. Devices that help maintain the quality of fuel and oil are known as purifiers. Purifiers may be of the disc or tubular type. These purifiers differ in construction and method; however, both operate to remove water and dirt from fuel by centrifugal force. As you can see from the information, you have read in this chapter, a great deal of importance is placed on keeping the fuel system in a high state of purity.

From reading our discussion of fuel injection systems, you should be aware of three types of mechanical diesel fuel injection systems: (1) *jerk type,* (2) *distributor-type*, and (3) *unit fuel injector type*, and how these three systems differ in construction and the methods by which fuel injection is achieved.

Our discussion of fuel injection would not have been complete without an explanation on how the amount of fuel delivered to the cylinders is controlled. The control of the engine speed is dependent on speed-sensitive devices known as governors. There are two basic types of governors, mechanical and hydraulic. The mechanical governor is used when extremely sensitive operation is not required. The hydraulic governors are more sensitive to speed variations. They are quicker acting and more accurate due to the reduction in friction of the mass of moving parts.

A complete understanding of fuel injection and engine control is necessary if operation of a diesel is to be in a safe and effective manner.

# CHAPTER 2

# DIESEL ENGINE OPERATING PRACTICES

In this chapter, we will apply the material in the preceding chapters to the practical problems of operating diesel engines. Since the diesel engines used by the Navy differ widely in design, size, and application, the procedures we will discuss apply only to general types of installations. Descriptions will apply generally to the various auxiliary and propulsion diesel installations in Navy ships. Detailed and specific information and operating instructions are provided in the manufacturers' manuals for specific installations, in NAVSEA technical manuals, and in ship's doctrine such as the Engineering Operational Sequencing System (EOSS).

After studying the information in this chapter, you should understand the purpose of EOSS and its value in contributing to effective engineering plant operations and casualty control. You should also be able to recognize the fundamental starting, operating, and stopping procedures for a diesel engine under normal operating conditions.

**Engineering Operational Sequencing System (EOSS)**

Each new class of ship that joins the Navy is more technically advanced and complex than the one before. This means that there is a need for more and better training of personnel who must keep the Navy's ships combat ready. The need for training and the problem of frequent turnover of trained personnel require the type of system that can be used to keep things going smoothly during the confusion of a casualty. The EOSS was developed for that purpose. The EOSS is a set of procedures and diagrams designed to eliminate problems because of operator error whenever personnel are aligning piping systems or starting and stopping the machinery.

The EOSS involves the participation of all personnel from the department head to the fireman on watch. The EOSS consists of detailed written procedures, with charts and diagrams, which were developed for safe operation and casualty control of a specific ship's engineering plant. The EOSS improves the operational readiness of the ship's engineering plant by providing positive control of the plant. In turn, this reduces operational casualties and extends machinery life.

The EOSS is divided into two subsystems: (1) Engineering Operational Procedures (EOPs) and (2) Engineering Operational Casualty Control (EOCC).

**Engineering Operational Procedures (EOPs)**

The EOPs are prepared specifically for each of the three following levels of operation:

1. Plant supervision (level 1)

2. Space supervision (level 2)

3. Component/system operator (level 3)

The materials for each level or stage of operation contain only the information necessary at that level. All materials are interrelated. Ship's personnel must use these materials together to maintain the proper relationship of each level of operation and to ensure positive control and sequencing of operational events within the plant. Personnel on ships that do not have EOSS should use written operating instructions and a casualty control manual for plant operations.

## Engineering Operational Casualty Control (EOCC)

The EOCC procedures enable plant and space supervisors to RECOGNIZE the symptoms of a possible casualty. They can then ELIMINATE or CONTROL the casualty to prevent possible damage to machinery and to RESTORE plant operation to normal. The EOCC contains procedures and information that describe symptoms, causes, and actions to be taken in the most common engineering plant casualties. We will discuss casualty control in more detail later in this chapter.

## Watch Standing

As a diesel engineers, they spend much of their time aboard ship as a watch stander. How they perform their watch is very important to the reliability of the engineering plant and the entire ship. They must have the skills to detect unusual noises, vibrations, or odors, which may indicate faulty machinery operation. They must also take appropriate and prompt corrective measures. They must be ready, at all times, to act quickly and independently. They must know the ship's piping systems and HOW, WHERE, and WHY they are controlled. They must know each piece of machinery; how it is constructed, how it operates, how it fits into the engineering plant, and where related equipment is controlled. They must read and interpret measuring instruments. They must understand how and why protective devices function (relief valves, speed-limiting governors, overspeed trips, and alarms). They must recognize and remove fire hazards, stow gear that is adrift, and keep deck plates clean and dry. They must NEVER try to operate a piece of equipment that is defective. They must report all unsafe conditions to the space and/or plant supervisor.

Whatever their watch station, they must know the status of every piece of machinery at their station. They must be sure that the logs are up to date and the status boards are correct, and they must know what machinery is operating before they relieve the watch. Above all, if they do not know—THEY ASK. A noise, odor, or condition may seem abnormal to one engineer, but he may not be certain whether it is a problem. Whenever they suspect any type of abnormal condition, they call their immediate watch supervisor.

They best gain the respect and confidence of their supervisors and shipmates when they stand an alert watch. Professional practices include relieving the watch on time or even a little early if possible and being sure they know the condition of the machinery and what they need to do. They must not *try to relieve the watch first and determine the details later.* The same rules should apply when they are being relieved. They should not be in a big hurry to run off. They should make certain their relief understands the situation completely. Before they are relieved, they must ensure their station is clean and organized (or as they would say, "squared away." These little considerations will earns the engineer a good reputation and improves the overall quality of watch standing within the department.

**Inspection and Maintenance**

Inspection and maintenance are vital to successful casualty control; they minimize casualties caused by material failures. Through continuous and detailed inspection procedures, damaged parts can be discovered, which may fail at a critical time, and eliminate underlying conditions, which will lead to early failure of parts. Underlying conditions will generally include maladjustment, improper lubrication, corrosion, erosion, and other causes of machinery damage.

Diesel engineers must pay particular and continuous attention to the following symptoms of malfunctioning equipment:

1. Unusual noises

2. Vibrations

3. Abnormal temperatures

4. Abnormal pressures

5. Abnormal operating speeds

They must thoroughly familiarize themselves with the specific temperatures, pressures, and operating speeds of equipment required for normal operation so that they will detect any departure from normal operation.

If a gauge or other instrument for recording operating conditions of machinery gives an abnormal reading, they must fully investigate the cause. The installation of a spare instrument or a calibration test will quickly indicate whether the abnormal reading is from instrument error. They must trace any other cause to its source.

Because of the safety factor commonly incorporated in pumps and similar equipment, considerable loss of capacity can occur before any external symptoms are apparent. Diesel engineers should be suspicious of any changes in the operating speeds (those normal for the existing load) of pressure-governor-controlled equipment. Variations from normal pressures, lubricating oil temperatures, and system pressures often indicate either improper operation or poor condition of machinery.

When a material failure occurs in any unit, promptly inspect all similar units to determine whether there is any danger that a similar failure might occur. Prompt inspection may eliminate a wave of repeated casualties.

They pay strict attention to the proper lubrication of all equipment, including frequent inspection and sampling to determine that the correct quantity of the proper lubricant is in the unit. It is good practice to make a daily check of samples of lubricating oil in all auxiliaries. Allow samples to stand long enough for any water to settle. When auxiliaries have been idle for several hours (particularly overnight), they should drain a sufficient sample from the lowest part of the oil sump to remove all settled water. Replenish with fresh oil to the normal level. Symptoms that indicate trouble may be in the form of an unusual noise or instrument indication, smoke, excessive consumption of lube oil, or contamination of the lube oil, fuel, or engine coolant.

**Symptoms of Engine Trouble**

When learning to recognize the symptoms that may help diesel engineers locate the causes of engine trouble, they will find that experience is the best teacher. Even though written instructions are essential for efficient troubleshooting, the information usually given will serve them only as a guide. It is very difficult to describe the sensation they should feel when they are checking the temperature of a bearing by hand; the specific color of exhaust smoke when pistons and rings are worn excessively; and, for some engines, the sound they will hear if the crankshaft counterweights come loose. They must actually work with the equipment before they can associate a particular symptom with a particular problem. Written information, however, can save them a great deal of time and can help them eliminate unnecessary work. Written instructions will make their detection of problems much easier in practical situations.

Symptoms that indicate trouble may be in the form of an unusual noise or instrument indication, smoke, excessive consumption of lube oil, or contamination of the lube oil, fuel, or engine coolant

**Noises**

Unusual noises that may indicate that a trouble exists (or is impending) are classified as pounding, knocking, clicking, and rattling. Diesel engineers must be able to associate each type of noise with certain engine parts or systems that might be the source of the trouble.

Pounding or hammering is a mechanical knock (and should not be confused with a fuel knock). Pounding or hammering may be caused by a loose, excessively worn, or broken engine part. Generally, troubles of this nature will require major repairs.

Detonation (knocking) is caused by the presence of fuel or lubricating oil in the air charge of the cylinders during the compression stroke. Excessive cylinder pressures accompany detonation. If detonation is occurring in one or more cylinders, they should stop the engine immediately to prevent possible damage.

**Symptoms of engine trouble**

Clicking noises are generally associated with an improperly functioning valve mechanism or timing gear. If the cylinder or valve mechanism is the source of metallic clicking, the trouble may be due to a loose valve stem and guide, insufficient or excessive valve tappet clearances, a loose cam follower or guide, broken valve springs, or a valve that is stuck open. A clicking in the timing gear usually indicates that there are some damaged or broken gear teeth.

Rattling noises are generally caused by vibration of loose engine parts. However, an improperly functioning vibration damper, a failed antifriction bearing, or a gear-type pump that is operating without prime are also possible sources of trouble when rattling noises occur.

When you hear a noise, first make sure that it is a symptom of trouble. Each diesel engine has a characteristic noise at any specific speed and load. The noise will change with a change in speed or load. As operators, diesel engineers must become familiar with the normal sounds of an engine. They must investigate all abnormal sounds promptly. They can detect and locate knocks that indicate trouble by using special instruments, such as an engineer's stethoscope, or a "sounding bar," such as a solid metal screwdriver or bar.

## Instrument Indicators

An engine operator probably relies more on the instruments to detect impending troubles than on all the other trouble symptoms combined. Regardless of the type of instruments you use, the indications are of no value if inaccuracies exist. Be sure an instrument is accurate and is operating properly. All instruments must be tested at specified intervals or whenever they are suspected of being inaccurate.

## Smoke

Smoke is a useful aid for locating some types of trouble, especially if you associate smoke in conjunction with other trouble symptoms.

The color of exhaust smoke can also provide you with clues in troubleshooting. The color of engine exhaust is a good, general indication of engine performance. The exhaust of an engine that is in good condition and operating under normal load has little or no color. A dark, smoky exhaust at normal load indicates incomplete combustion; the darker the color, the greater the amount of unburned fuel in the exhaust. Incomplete combustion may be due to a number of troubles. Some manufacturers associate a particular type of trouble with the color of the exhaust. The more serious troubles are identified below:

1. Bluish-white smoke

   a. Worn or stuck piston rings
   b. Worn cylinder liners
   c. Worn valve guides
   d. Cracked pistons
   e. Leaking injectors

2. Black or gray smoke

   a. Incompletely burned fuel
   b. High exhaust backpressure (clogged exhaust ports, piping, or muffler)
   c. Restricted air inlet (clogged inlet ports, air cleaner, and blower inlet screen)
   d. Malfunctioning turbocharger
   e. Improperly timed or faulty injectors
   f. Engine overload (cylinders not balanced)
   g. Low compression (burned valves or stuck piston rings)

## Excessive Consumption of Lube Oil, Fuel, or Water

An operator should be aware of engine trouble whenever excessive consumption of any of these vital liquids occurs. The possible troubles indicated by excessive consumption will depend on the system in question. Leakage, however, is one trouble that may be common to all systems. Before starting any disassembly, check for a misaligned system or for leaks in the system in which any excessive consumption is occurring.

## Electrical Systems

Since most Navy small boat crews do not include an electrician, it will be the responsibility of the boat engineer to troubleshoot and repair any problems in the electrical ignition and lighting systems. The ignition systems have been discussed earlier in this manual, so most of the information here will apply to auxiliary systems. The electrical system on a typical small boat consists of the following equipment and devices:

1. A battery-charging generator or alternator driven by the propulsion engine

2. An engine starting system

3. A battery used both for starting the engine and supplying auxiliary loads when secured

4. A control and distribution panel having switches and fuses for the control and protection of circuits to auxiliary loads

5. Cables for interconnecting the above

6. A voltage regulator

## Ungrounded, Two-Wire, 24-Volt System

The electrical system used on Navy small boats is generally a two-wire, ungrounded, 24-volt system. The two-wire system is a necessity on a non-conducting boat hull of wood or plastic. The steel boat hull, like the automobile chassis, makes a good electrical conductor and permits the use of a single-wire electrical system with a grounded or hull return. There are certain reasons, however, for the use of an ungrounded, two-wire system on steel hull boats instead of the single-wire installation.

Because of environmental conditions (such as exposure to saltwater spray), an unwanted ground sometimes occurs on boat electrical systems. Experience has shown that fewer shutdowns occur on ungrounded, two-wire systems than on the single-wire systems. As a result, every effort has been made to provide a two-wire, ungrounded system, and the Electrical Engineer (or Electrician's Mate) must maintain that system in good condition. An ungrounded system can tolerate the temporary situation of any SINGLE grounded condition, regardless of its location, because no function is affected. A variety of troubles, however, may result when TWO places in the same system become grounded. The more common troubles include blown fuses, failure of the starting system to energize, or faulty operation of any device, such as horns, voltage regulator, lights, and miscellaneous auxiliary loads.

**Protection of Circuits**

Fuses are normally provided only in the circuits supplying auxiliary loads, such as horns, running lights, cabin lights, spotlights, and communications equipment.

All other circuits, such as the starting motor circuit, solenoid switch control circuit, battery charging circuit, and power supply to the distribution panel, are unfused. This is because the possibility of short circuits or leakage currents is reduced by use of the following equipment, components, and systems:

1. Two-wire, ungrounded electrical system instead of single-wire, grounded

2. Two-wire, ungrounded electrical components of such rugged construction as to make grounding of internal wiring or terminals difficult under normal service conditions

3. Watertight components, such as battery connection boxes, a distribution and control panel, and a starting motor solenoid switch

4. Cables between batteries and starting motor of sufficient size to carry high inrush currents and provided with terminal lugs and end sealing to prevent penetration of moisture

5. Splash guards for attached generators/ alternators

6. Sealed meters or transparent splash shields to protect instruments such as battery charging ammeters

7. Fusing of auxiliary load circuits so that faults in these circuits can be isolated

When diesel engineers are working around an engine or boat, it will be to their advantage to pay particular attention to how a unit is wired and what type of wiring is used. These observations will make their job of finding and repairing troubles much easier.

When they are on a small boat away from the ship, it will be rather difficult for you to check out a system completely because not all the required test equipment will be available. However, they can check for such things as loose or corroded connections, broken wiring, faulty switches, and burned out lamps or fuses.

## Operating Instructions for Diesel Engines

There may be occasions when a diesel engine must be started, operated, and secured under a variety of demanding conditions, such as emergencies and casualties in engine supporting systems. Operation under such unusual conditions requires that diesel engineers know and understand the engine installation, the function of supporting systems, and the reasons for the procedures used in normal and emergency operations.

The procedures we will discuss in the following sections are basic steps of engine operation. These procedures do NOT contain every step that must be taken. Remember, circumstances and conditions concerning engine operation will vary. When diesel engineers you are starting or operating an engine or combating casualties in the engineering plant, use your EOSS.

## Starting Procedures

Diesel engines are started either by hydraulic, electric, compressed air admission, or by air-powered starting motors. The general starting procedure for all types of systems consists of (1) making pre-operational checks, (2) aligning supporting systems, and (3) cranking the engine with the starting equipment until ignition occurs and the engine is running.

The steps of the starting procedure will differ depending on whether they are starting the engine after routine securing, after a brief period of idleness, or after a long period of idleness. We will first list the basic steps they should follow for starting an engine that has been routinely secured under normal conditions.

## After Routine Securing

To start an engine that has been routinely secured, you should first make ready the supporting systems—cooling, lubrication, and fuel—as follows:

1. Check all valves in the seawater cooling system to ensure that the system is lined up for normal operation.

2. Start the separate motor-driven seawater pump (if it is provided). If an auxiliary engine is cooled from the ship's seawater circulating system, ensure that adequate pressure and flow will be available.

3. Vent seawater coolers, using the vent cocks or vent valves on the heat exchanger shells. (If this is not done, air or gas can accumulate, reducing the effective cooling surface area of a heat exchanger.)

4. Check the level in the freshwater expansion tank. Remember that a cold expansion tank will need a lower fluid level than one that is hot, so leave room for expansion.

5. Check the freshwater cooling system: Set all valves in their operating positions, start the motor-driven circulating pump (if it is provided), vent the system, and check the freshwater level in the expansion tank again. The freshwater level may have dropped if air or gas were vented elsewhere from the system.

6. Check the lubricating system: Check the oil level in the sump; add oil if necessary to bring it to the proper level. Ensure that adequate

grease is applied to bearings that require grease lubrication. If oil sump heaters are installed, raise the lubricating oil temperature to 100°F.

7. In idle engines, the lube oil film can be lost from the cylinder walls. It is desirable for you to restore this film before you actually start the engine. (Large diesel engines will restore the film by pressurizing the lube oil system and jacking the engine over without starting it. The pressure in the lube oil system will oil the cylinders, and the pistons will distribute the oil film.) To pressurize the lubricating system, either start the motor-driven lubricating oil pump, or air-driven pre-lube pump (if installed), or operate a hand operated lubricating oil pump. If the engine drives the lubricating oil pump, it will develop pressure when the engine is jacked over. To reduce the load on the jacking gear and prevent an accidental start, open any cylinder test valves or indicator cocks. Then turn the engine over using the jacking gear, which may be motor-driven or hand-operated. As the engine turns over, observe the indicator cocks for excessive moisture. The presence of excessive moisture indicates water or fuel accumulation in the cylinders.

8. When the diesel engineer has performed the preceding operation, he will disengage the jacking gear and restore the cylinder test valves or indicator cocks to their operating positions.

9. Line up and prime the fuel systems. Check to ensure that there is sufficient clean fuel for the anticipated engine operation.

10. Test the alarm panel for power by manually operating such alarms as the low pressure lubricating oil alarm and the freshwater high-temperature alarm.

11. Now start the engine with the starting system. Follow the approved written procedures for the type of starting system in use.

12. Once the engine is running, energize the low-pressure lube oil alarm and the water temperature alarm. Pay careful attention to all gauges and other indications of engine condition and performance. Diesel engines tend to be noisy, particularly when they are cold and idling. Familiarity with the normal sounds of the engine will help avoid unnecessary panic.

    If the lube oil pressure does not rise immediately to the operating pressure, STOP the engine and determine the cause of the low pressure.

13. Idle the engine until the lube oil temperature reaches 100°F. Next, apply a light load of 20 to 30 percent. When the lube oil temperature reaches 120°F, apply the normal load (50 to 80 percent). If possible, avoid placing a load on the engine until the engine has reached operating temperature. Normal or high loading of a cold engine will produce carbon in the cylinder heads, cause excessive engine wear, and dilute the lubricating oil. The procedures for placing the engine "on the line" will depend on the type of installation. In general, it is best to bring the engine up to speed gradually, while being alert for symptoms of trouble when you are initially loading the engine and while the engine is approaching normal range.

## After a Brief Period of Idleness

Starting a warm engine, after it was recently secured and if no unusual conditions are suspected, consists of (1) aligning the systems that may have been secured (such as circulating water), (2) disconnecting the engine from the load, and (3) cranking the engine up to starting speed. Carefully observe the lubricating oil pressure. The temperature of coolant may exceed normal operating temperatures for a minute or so until the heat accumulated in the secured engine is removed.

## After Overhaul or a Long Period of Idleness

Diesel engineers make additional checks and inspections when the engine being started has been idle for a long period or has been recently overhauled. The following checks should be performed:

1. Inspect the parts of the engine system that have been worked on to ensure that the work is complete, that the covers have been replaced, and that it is safe to operate any valves or equipment that have been tagged out of service. (All DANGER tags must be removed.)

2. Check all pipe connections to see whether the connections are tight and whether the systems have been properly connected.

3. Fill the freshwater cooling system with treated water if it has been drained. Be sure coolant is flowing through all parts and components of the system. Vent the system.

4. Make a thorough check of the lubricating system. Check the sump level and fill the sump if necessary. If a separate oil pump is installed, pre-lube the engine. You can consider the system to be pre-lubed when a slight pressure registers on the engine oil pressure gauge. Then make a visual check, with inspection plates removed, to see whether oil is present at all points of the system and in each main bearing. Examine pipes and fittings for leaks. If lubricators are installed, be sure they are filled.

5. Inspect the air receiver, the filter, and the discharge passages of the blower for cleanliness, and remove any oil accumulations.

6. If the engine has a hydraulic governor, inspect the governor oil level. If an overspeed trip is installed, be sure it is in proper operating condition and position.

7. Examine all moving parts of the engine to see that they are clear for running. Check the valve assemblies, including the intake, exhaust, and air-starting valves, and the fuel control linkage for freedom of movement.

8. Inspect the fuel service tank for the presence of water and sediment. Fill the tank with clean fuel if necessary. Start the auxiliary fuel pump (if one is installed) and see whether the fuel pressure gauges are registering properly. Examine the fuel piping and fittings for leaks, especially the fittings and lines inside the engine. Thoroughly vent all air from the fuel system, using the vent cocks. Be sure that the fuel strainers have been cleaned and that new filter elements have been installed.

9. If the engine has an air-starting system, open the lines on the system and blow them out. Reconnect the lines and pressurize the starting air banks (flasks).

10. Make a final check to ensure that all parts are in place, then open all scavenging-air header and exhaust header manifold drains.

11. Now start the engine, using the procedures for a routinely secured engine.

**Normal Operating Procedures**

Operation of a diesel engine cannot be separated from the operation of the equipment the engine is driving. Therefore, for purposes of our discussion, we will assume that operators are fully aware of the complete system you are running. Each type of engine and installation has its special operating routine. A systematic procedure has already been established, based on these special requirements and on the experience of the engine operators with the particular installation. They must respect and follow this procedure.

The following information is general and should be considered as incomplete in terms of operation of any specific plant.

While an engine is operating, its performance is monitored and observed for two purposes: (1) to recognize early any unsatisfactory operation or impending malfunctions so that immediate casualty control procedures can be started, and (2) to develop a comparative record over a period of time so that gradually deteriorating conditions can be detected. For the latter purpose, engineers must keep a complete log of all operating conditions. Observe and record the operating pressures and temperatures in the log at hourly intervals. Compare the entries over a period and note any deviations from normal conditions.

Diesel engineers must be alert to changing or unusual noises made by the operating machinery. Gradually changing sounds are difficult to detect, especially if inexperienced. Often an oncoming watch will detect a new sound that the present watch was unaware.

When unusual operating conditions occur, load, lubrication, cooling, engine speed, or fuel supply problems are usually responsible, either directly or indirectly. You must be alert to changes in any of these areas. In the next sections, we will provide you with some general guidelines.

## Load

The manner of applying a load to an engine and the regulation of the load will depend on the type of load and the design of the system. The procedures for loading an engine, or placing it on the line, are established by the EOSS.

Whenever you are starting a cold engine, allow ample time to build the load up gradually. NEVER fully load an engine before it has been warmed up. Gradual application of the load will prevent damage to the engine from such conditions as uneven rates of expansion and inadequate lubrication at low temperatures. (An exception to this rule is the use of emergency generators, set up for automatic mode, in which the engines must take on rated loads as soon as they are started.)

Never operate a diesel engine for prolonged periods with less than one-third of its rated load. Combustion at low load is incomplete, so partially burned fuel and lubricating oil may cause heavy carbon deposits, which will foul the valve stems, injector tips, piston rings, and exhaust systems.

In addition to these problems, prolonged operation at low-load conditions may cause the exhaust valves to stick and burn, dilute the lubricating oil, scuff the cylinder liners, increase fuel consumption, and cause excessive smoke when the load is increased. If an engine must be operated at less than 30 percent power for more than 30 minutes, the load should be increased to above 50 percent power at the first opportunity.

Diesel engines are designed to operate up to full-load conditions for prolonged periods. However, diesel engines should NEVER be operated at an overload except in an emergency. This includes both excessive torque and engine speed. Excessive firing pressures and exhaust temperatures may indicate overload. When conditions indicate an overload, reduce the load immediately.

**Lubrication**

The performance of the lubrication system is one of the most important factors of engine operation you can monitor. Indicators continuously show oil temperature and pressure in key parts of the system. While the engine is operating, the indicators and sight glasses should be monitored on a regular basis. An alarm (horn or siren) will usually warn of low pressure. If an alarm is not installed (as for example on an engine that is used in a small boat), diesel engineers must continuously monitor the oil pressure and check the oil level. Under typical operating conditions, they should be able to estimate the rate at which the engine burns its lubricating oil and to predict when replenishment will be needed.

The condition and cleanliness of the lubricating oil is critical for long engine life. Therefore, they should clear the metal-edge type of lubricating oil strainers by rotating the cleaning handle, which should be performed during each watch. The condition of filters is often indicated by the amount of pressure drop from the inlet to the outlet. Gauges are installed to indicate this differential. These gauges should be checked frequently. As specified by PMS, a test kit is available for them to use to check the condition of the lubricating oil.

**Pressures and Temperatures**

Ensure that all pressures and temperatures are maintained within the normal operating ranges. If this is not possible, secure the engine. Check all instruments frequently. The NAVSEA technical manual provides detailed information concerning the proper operating pressures and temperatures. When this information is not available, maintain the temperature of the lubricating oil as it leaves the engine between 160° and 200°F (180° F is preferred), and maintain the temperature of the fresh water as it leaves the engine at not less than 155°F or more than 185°F (170°F is preferred). Do not allow the temperatures in the seawater cooling system to exceed 130°F. Higher temperatures will cause deposits of salt and other solids in the coolers and piping and will aggravate corrosion.

Make frequent checks of the cooling system to detect any leaks. Vent coolers and heat exchangers at least once each watch. Check the level of the fresh water in the expansion tank frequently and add fresh water as necessary. If the freshwater level gets low enough to cause overheating of the engine, NEVER add cold water until after the engine has cooled.

## Critical Speeds

The vibrations resulting from operation at destructive critical speeds will cause serious damage to an engine.

All moving parts of machinery have critical speeds. *Critical Speed* means there are certain ranges of speed during which excessive vibration in the engine will be created. Every part of the engine has a natural period of vibration and frequency. When impulses set up a vibration that coincides with the natural frequency of the body, each impulse adds to the magnitude of the previous vibration. Finally, the vibration becomes great enough to damage the engine structure.

Vibration may be set up by linear impulses from reciprocating parts or by torsional impulses from rotating members. The crankshaft is the part that causes torsional vibration, because pressure impulses on the piston cause the crankshaft to twist. When the pressure acting on the piston in each cylinder decreases, the shaft untwists. If pressure impulses, which are timed to the natural period of the shaft, are permitted to continue, the amplitude of vibration will become so great that the shaft can break. If the speed of such an engine is changed, however, the pressure impulses will no longer coincide with the natural period of the shaft and the excessive vibration will stop.

Since each engine has a natural period of vibration (which cannot be changed by the operator), the only control you have is to avoid operating the engine at critical speeds. If critical speeds exist below the normal speed of the engine, pass through the critical ranges as quickly as possible when changing engine speed. Detailed information concerning critical speed ranges is provided with each installation. Tachometers should be marked to show any critical speed ranges to make it easier to keep the engine out of the critical ranges. Remember, tachometers sometimes get out of adjustment. Consequently, frequently compare each tachometer with a calibrated mechanical counter.

## Fuel

Maintain an adequate supply of clean fuel. Check the fuel system frequently for leaks. Clean all fuel strainers at periodic intervals. Replace fuel filter elements whenever necessary. When diesel fuel purifiers are provided, purify all fuel before transferring it to the service or day tanks. Frequently check the service tanks for water and other settled impurities by sampling through the drain valve at the bottom of the tank. Drain off water and impurities.

## Stopping and Securing Procedures

A diesel engine can be stopped by shutting off the fuel or air supply. The fuel supply can be shut off by placing the throttle or the throttle control in the STOP position. If the engine installation permits, it is a good idea to allow the engine to idle, without load, for a short time before stopping it. This practice will permit engine temperature to reduce gradually. It is also good practice to operate the manual overspeed trip when stopping the engine so that the operating condition of the device can be checked. Before tripping the overspeed trip, reduce the engine speed to the specified idling speed. Some overspeed trips reset automatically. In some installations, however, the overspeed trip must be reset manually before the engine can be started again.

In addition to the detailed procedures listed in various EOSS checklists and NAVSEA technical manuals, take the following steps after an engine has stopped:

1. Open the drain cocks on the exhaust lines and on the scavenging-air inlet headers (if they are provided).

2. Leave open the specified number of indicator cocks, cylinder test valves, or hand operated relief valves to detect any water accumulation in the cylinders prior to starting the engine.

3. Secure the air pressure. If starting air is left on, the possibility of a serious accident will increase.

4. Close all sea valves.

5. Allow the engine to cool.

6. Clean the engine thoroughly by wiping it down before it cools. Clean the deck plates and see that the bilges are dry.

7. Arrange to have any casualties repaired. No matter how minor casualties may appear, repairs must be made and troubles must be corrected promptly.

**Precautions in Diesel Engine Operation**

Obtain the specific safety precautions for a given engine from the appropriate NAVSEA technical manual. In addition to the guidelines in the NAVSEA technical manual, observe the following precautions when operating or maintaining a diesel engine.

**Relief Valves**

If a relief valve on an engine cylinder lifts (pops) several times, stop the engine immediately. Determine the cause of the trouble and decide upon the correct solution. Except in an emergency, NEVER lock a relief valve in the closed position. Pressure-relief mechanisms are fitted on enclosures in which excessive pressures may develop.

**Fuel**

When fuel reaches the injection system, it should be absolutely free of water and foreign matter. Thoroughly centrifuge the fuel before using it, and keep the filters clean and intact. Remember, fuel leakage into the lubricating oil system will cause dilution of the lubricating oil with a consequent reduction in viscosity and lubricating properties.

## Cooling Water

Do NOT allow a large amount of cold water, under any circumstances, to enter a hot engine suddenly. Rapid cooling may crack a cylinder liner and head or may cause a piston to seize within a cylinder. Reduce the load or, when ordered to do so, stop the engine when the volume of circulating water cannot be increased and the temperatures are too high. In freezing weather, you must carefully drain all passages and pockets in the engine that contain fresh water and that are subject to freezing, unless an antifreeze solution has been added to the water.

## Starting Air

When engines are stopped, vent all starting-air lines. Serious accidents may result if pressure is left on. Intake air must be kept as clean as possible. Accordingly, keep all air ducts and passages clean.

## Cleanliness

Cleanliness is essential to efficient operation and maintenance of diesel engines. Maintain clean fuel, clean coolants, clean lubricants, and a clear exhaust. Keep the engines clean at all times, and take steps to prevent oil or fuel from accumulating in the bilges or in other areas to prevent fire hazards.

## Emergency Diesel Generators

Most naval ships are equipped with diesel driven emergency generators (diesel engines are most suitable for this application because of their quick-starting ability). Emergency generators furnish power directly to vital electrical auxiliaries, such as the steering gear and the ship's gyro. In addition, emergency generators may serve as a source of power for the casualty power distribution system. All engineering personnel should become familiar with the emergency and casualty power systems aboard their ship.

Emergency diesel generator sets must be ready at all times for immediate use. Complete the following checks to ensure that the support systems and control system are aligned and that the emergency generator is ready for operation.

1. Fuel service tank filled, with all water drained.

2. Fuel system valves correctly aligned.

3. Air flask(s) charged

4. Air-starting system valves correctly aligned

5. "Keep warm" system(s) (if used) activated

6. Switchboard set to AUTOMATIC position.

A typical shipboard plant may consist of two emergency diesel generators, one forward (near the bow and above the waterline) and one aft (near the stern), in spaces outside the main machinery spaces. Each emergency generator has its individual switchboard and switching arrangement for control of the generator and for distribution of power to certain vital auxiliaries and to a minimum number of lighting fixtures in vital spaces.

The capacity of the emergency unit varies with the size of the ship in which it is installed. Regardless of the size of the installation, the principle of operation of the engine is basically the same as it is for any diesel engine.

Emergency diesel engines are started either by compressed air or by a starting motor and develop full-rated load power within 10 seconds of starting. In a typical installation, the starting mechanism is actuated when the ship's normal supply voltage on the bus falls to approximately 80 percent. (In a 440-volt system, this would be approximately 350 volts.) The generators are not designed for parallel operation. Therefore, when the ship's supply voltage fails, a transfer switch automatically disconnects the emergency switchboard from the main distribution switchboard and connects the emergency generator to the emergency switchboard. With this arrangement, transfer from the emergency switchboard back to the main distribution switchboard is accomplished manually. Then, the emergency generator must be manually stopped and reset for automatic starting.

Since emergency diesel generators are of limited capacity, only certain circuits can be supplied from the emergency bus. These include such circuits as the steering gear and the interior communications switchboard. If some vital circuit is secured, another circuit may then be cut in, up to the capacity of the generator.

## Operating Instructions

Normally, the emergency diesel generator will start automatically, but for test purposes and under other conditions it may be started and operated manually. The following guidelines are for testing the operation of an air-started emergency diesel generator set.

The engine is started automatically when the ship's normal supply fails and causes the solenoid air valve (located between the starting-air flask and the engine) to open, admitting starting air to the engine. The engine then turns over on air until firing begins. As the engine speed increases, the air cutoff governor valve closes and shuts off the starting air. As soon as the normal operating speed is reached and the generator develops normal voltage, the solenoid air valve also closes to shut off the starting-air supply. (The starting air flask is charged from the high-pressure air system, through a reducing valve. The air stored in the starting-air flask varies in pressure from 300 to 600 psi, depending on the installation.)

To start the engine manually, de-energize the solenoid valve. If the ship's supply current is not broken, you must open the switch in the solenoid circuit. Then, admit starting air to the engine by opening the valve manually with the hand wheel. After firing begins, turn the hand wheel to close the valve and cut off the starting air. The hand wheel must be turned to the open position of the valve whenever you must leave the generator set available for emergency service.

If the lubricating oil pressure does not build up immediately after the engine starts, shut down the engine and determine the cause of the trouble. NEVER operate the engine without lubricating oil pressure. At regular intervals, check the lube oil pressure, fuel pressure, cooling water temperature, and exhaust temperature. In addition, clean the fuel and lubricating oil filters regularly.

To SHUT DOWN or STOP the engine, move the fuel-control lever to the STOP position. After the lever is released, it will automatically return to the running position to permit the engine to be restarted.

**Operating Precautions**

Observe the following operating precautions and inspections:

1. Do NOT operate the engine without lubricating oil pressure; this will cause serious damage.

2. Do NOT operate the engine in an overloaded or unbalanced condition. An overload condition on one or more cylinders may be indicated by an increase in the exhaust temperature or by smoky exhaust.

3. Do NOT operate the engine with an abnormal water outlet temperature.

4. Do NOT operate the engine after an unusual noise develops; the noise might be an indication of pending trouble. Investigate the noise and correct any trouble, particularly if the condition may prove harmful to the engine.

5. If the overspeed device trips and shuts down the engine, investigate the cause of the trouble before restarting the engine.

6. Make certain that the fittings of the ventilation system that serve the compartment in which the engine is located are open. If a diesel engine is started while the vent system is secured, the engine will consume the air in the compartment. Under these conditions, the engine may continue to operate long enough for suffocation. This precaution applies to installations where the engine does not have a direct air supply from the outside to the intake manifold. THESE PRECAUTIONS ALSO APPLY TO EMERGENCY DIESEL FIRE PUMPS.

## Factors Influencing Engineering Casualty Control

The scope of engineering casualty control is much broader than the immediate actions that are taken at the time of a casualty. Engineering casualty control reaches peak efficiency through a combination of sound design, careful inspection, thorough plant maintenance (including preventive maintenance), and effective personnel organization, management, and training. CASUALTY PREVENTION IS THE MOST EFFECTIVE FORM OF CASUALTY CONTROL.

The primary instructions and guidelines used to handle any engineering casualty aboard Navy ships are as follows:

1. Engineering Operational Casualty Control (EOCC) procedures

2. Ship's casualty control manual (for ships without EOCC)

3. Ship's damage control manual

4. Ship's damage control bills (part of the ship's Watch, Quarter, and Station Bill)

5. Ship's Organization and Regulations Manual (SORM)

**Design**

Sound design influences the effectiveness of casualty control in two ways: (1) it eliminates weaknesses, which may lead to material failure, and (2) it installs alternate or standby means for supplying vital services in the event of a casualty to the primary means. Both of these factors are considered in the design of naval ships. Individual plants on board ship are equipped with duplicate vital auxiliaries, loop systems, and cross-connections. Complete propulsion plants are also designed to operate as isolated units (split-plant design).

**Communications**

Casualty control communication is vitally important to the operation and organization of the ship. Without adequate and proper means of communication, the whole organization of casualty control will fail in its primary objective.

As a provision for sufficient means of communication to be available, several different systems are installed aboard ship. The normal means of communications are the battle telephone (sound-powered) circuits, interstation 2-way systems (intercoms), ship's service telephones, ship's loudspeaker (1-MC), and voice tubes. Messengers are used in some situations when other methods of communication are not available or when written reports are required.

Transmission of correct information regarding a casualty and the speed with which the report is made are the principal values of any method of communication.

Control of all communication circuits must be established by the control station. The circuits must never be allowed to get out of control from "cross talk" caused by more than one station. Casualty control communications must be incorporated into casualty control training. The control station or engineering control must be promptly notified of a casualty so that other casualties (which could be more serious than the original casualty) can be prevented.

**Training**

Casualty control training must be a continuous systematic procedure with constant refresher drills. Realistic simulation of casualties requires adequate preparation. The amount of advance preparation required is not always readily apparent. You must carefully visualize the full consequences of any error that could be made in handling simulated casualties that were originally intended to be of a relatively minor nature. There must be a complete analysis and all participants must be carefully instructed before simulation of major casualties and battle damage. A new crew must have an opportunity to become familiar with the ship's piping systems and equipment before simulation of any casualty that may have other than purely local effects.

In the preliminary phases of training, a "dry run" is useful for imparting knowledge of casualty control procedures without endangering the ship's equipment by a too realistic simulation of a casualty. Under this procedure, a casualty is announced, and all individuals are required to report as though action were taken (an indication must be made that the action is simulated). Definite corrective actions can be taken, and with careful supervision, the timing of individual actions can appear to be very realistic. Regardless of the state of training, dry runs should always be held before actual simulation of any involved casualty. Similar rehearsals should be held before simulation of relatively simple casualties whenever new personnel are involved and particularly after an interruption (such as a naval shipyard overhaul period) of regularly conducted casualty training has occurred.

## Correction and Prevention of Casualties

The speed with which corrective action is taken to control an engineering casualty is of paramount importance. This is particularly true for casualties that affect the ship's propulsion power plant, steering system, and electrical power generation and distribution. If casualties associated with these functions are allowed to accumulate, they may lead to serious damage to the engineering installation—damage that often cannot be repaired without loss of the ship's operating availability. When risk of possible permanent damage exists, the commanding officer has the responsibility of deciding whether to continue operation of the equipment under casualty conditions. Such action can be justified only when the risk of even greater damage, or loss of the ship, may be incurred if the affected unit is immediately secured.

Whenever there is no probability of greater risk, the proper procedure is to secure the malfunctioning unit as quickly as possible even though considerable disturbance to the ship's operations may occur. Although speed in controlling a casualty is essential, action should never be undertaken without accurate information; otherwise, the casualty may be mishandled and cause irreparable damage and possible loss of the ship. War experience has shown that the cross-connecting of intact systems with a partly damaged one must be delayed until it is certain that such action will not jeopardize the intact systems. Speed in handling casualties can be achieved only by thorough knowledge of the equipment and associated systems and by thorough and repeated practice in performing the routines required to control specific, predictable casualties.

PHASES OF CASUALTY CONTROL

The handling of any casualty by shipboard personnel can usually be divided into three phases:

(1) limiting of the effects of the damage,

(2) emergency restoration, and

(3) complete repair.

The first phase is concerned with immediate control of a casualty to prevent further damage to the affected unit and to prevent the casualty from spreading through secondary effects, commonly known as "cascading." (One fault leads to another.)

The second phase requires the use of Engineering Operational Procedures (EOPs) and involves restoring, as far as practicable, the services that were interrupted as a result of a casualty. For many casualties, the completion of this phase will eliminate all operational handicaps, except for the temporary loss of standby units, which will lessen the ability of the machinery to withstand additional failure. If no damage to, or failure of, machinery has occurred, this phase usually completes the operation.

The third phase of casualty control consists of making any repairs required to completely restore the installation to its original condition.

**Split-Plant Operation**

A primary method of casualty prevention and control is use of the split-plant mode of operation. The purpose of the split-plant design is to minimize battle damage that might result from a single hit.

Most naval ships that were built primarily as warships have at least two engineering plants. Larger combatant ships have four individual engineering plants. Split-plant operation means aligning support systems, engines, pumps, and other machinery so that two or more propulsion plans and/or electrical generating plants are available, each complete in itself. Each main engine installation has its own piping systems and other auxiliaries. Each propulsion plant operates its own propeller shaft. If one plant were to be put out of action by explosion, shellfire, or flooding, the other plant could continue to drive the ship ahead, though at somewhat reduced speed.

Split-plant operation is not absolute insurance against damage that might immobilize the entire engineering plant, but it does reduce the chances of such a casualty. It prevents transmission of damage from one plant to another or possible serious effect on the operation of the other plant or plants. It is the first step in the PREVENTION of major engineering casualties.

The fuel system is generally arranged so that fuel transfer pumps can take suction from any fuel tank in the ship and can pump to any other fuel tank. Fuel service pumps supply fuel from the service tanks to the main engines. In split-plant operations, the forward fuel service pumps of a ship are lined up with the forward service tanks, and the after service pumps are lined up with the after service tanks. The cross-connect valves in the fuel transfer line must be closed except when fuel is being transferred.

Diesel propulsion plants are designed for split plant operation only; however, some of the auxiliary and main systems may be run cross-connected or split. Among these auxiliaries are the starting-air systems, cooling-water systems, firemain systems, and, in some plants, the fuel and lube oil systems.

In diesel-electric installations, the diesel elements are split, but the generator elements can be run split or cross connected. The advantages of this type of installation will depend on operating procedures as well as design.

## Locking Main Shaft

An engineering casualty may be such that continued rotation of the main shaft will cause further damage. The main shaft should be locked until necessary repairs can be made since, except at very low speeds, movement of the ship through the water will cause the shaft to turn. Turning of the shaft will occur whether the ship is proceeding on its own power or being towed.

For locking a main shaft, there are no standard procedures applicable to all types of diesel-driven ships. For ships that have main reduction gears, shaft locking of the turning gear is permissible, provided it is designed for this purpose. Some ships have brakes that are used to hold the shaft stationary. On diesel-electric drive ships, no attempt should be made to hold the shaft stationary by energizing the electrical propulsion circuits.

## Emergency Procedures

Under certain circumstances, you may be ordered to start additional engines. Time may not permit following the normal, routine procedures. Emergency procedures may have to be used. Because emergency procedures will differ, depending on the installation, be familiar with the specific procedures established for the ship.

## Engine-Room Casualties

In the event of a casualty to a component of the propulsion plant, the principal objective is to prevent additional or major casualties. Where practicable, the propulsion plant should be kept in operation with standby pumps, auxiliary machinery, and piping systems. The important thing to remember is to prevent minor casualties from becoming major casualties, even if it means suspending the operation of the propulsion plant. It is better to stop the main engines for a few minutes than to risk putting them completely out of commission, a condition that will require major repairs.

When a casualty occurs, the engineering officer of the watch (EOOW) and the petty officer of the watch (POOW) must be notified immediately. The watch officer will notify the OOD and the engineer officer. Main engine control must keep the bridge informed as to the nature of the casualty, the ship's ability to answer bells, the maximum speed available, and the probable duration of the casualty.

**General Safety Precautions**

In addition to following the specific safety precautions listed in the operating instructions for an engine, continuously exercise good judgment and common sense when taking steps to prevent damage to material and injury to personnel. In general, help to prevent damage to machinery by operating engines according to prescribed instructions, by using practices such as "bagging and tagging" parts that were removed from an engine during maintenance or overhaul, by having a thorough knowledge of duties, and by being totally familiar with the parts and functions of the machinery being operated and maintained.

By maintaining machinery so that the engines will be ready for full-power service in the event of an emergency and by taking steps to prevent conditions that are likely to constitute fire or explosion hazards, you can also help to prevent any damage that might occur outside of the ship. This type of damage may take the form of damage to piers or other external structures or to other marine craft whenever a loss of control over the ship occurs.

Remember, personnel work most safely when they thoroughly know how to perform their duties, how to use their machines, how to take reasonable precautions around moving parts, and when they are consistently careful and thoughtful while performing their duties.

## Emergency Starting and Securing Procedures

There may be times when an engine must be started, operated, or secured under emergency conditions. Before this becomes necessary, operating personnel should learn the procedures in the ship's EOCC. These procedures should be posted at the engine control station or operating position. Operators should be drilled in casualty control procedures at regular intervals.

There is a definite hazard to starting a diesel engine under emergency conditions because personnel are rushed and tend to be careless. There is always time to ensure that personnel are clear of external moving parts, such as belt drives and shafts, before actuating the starting gear. If emergency repairs have been made, be sure that all tools are accounted for before you close up the engine and that all essential parts have been replaced before you start the engine. An engine can be started and run briefly if it has air and fuel and if the starting system will operate. It will run much longer if it has functional lubrication and cooling systems. With the exception of some boat engines that can be started by towing, there is no backup for the starting system. Usually sufficient spare parts and resources are available to restore any casualty to the starting system. Remember, however, if the repair is rushed, the danger resulting from careless work will increase.

In an emergency, start an engine by lining up the fuel system and actuating the starter. Before doing this, however, make certain that there is a supply of air to the engine and engine compartment and that the lubricating system will operate. After starting, establish cooling-water flow and review all the normal prestarting checks as quickly as possible.

If an operating engine suffers a casualty, the decision of whether to continue operating or to secure the unit must be made immediately. The condition of the ship's operation is an important factor in this decision. In some instances, when risk of possible permanent damage exists, the commanding officer has responsibility for deciding whether to continue operation of equipment under casualty conditions. Such action can only be justified when the risk of greater damage, or loss of the ship, may be incurred if the affected unit is secured. Risk to the ship is present in actual combat situations, severe weather conditions, narrow channels, and potential collision situations, which include close-formation maneuvering with other ships.

Engines can be operated with casualties to vital auxiliaries if the function of the auxiliary unit can be produced by other means. For instance, cooling-water flow can be reestablished from a firemain, and an engine can operate for some time with seawater in its cooling system as long as the cooling system is rinsed well afterward.

If the decision is made to secure an engine that has suffered a casualty, the general rule is to stop the engine as soon as possible. In the case of a propulsion engine, it will usually be necessary to stop the shaft also. This may require slowing the ship until the shaft is stopped and locked with the turning gear, shaft brake, or other means.

Often, the engine can be stopped by securing the flow of fuel. Occasionally, this method will not work since a blower seal leak or a similar situation may permit the engine to run on its own lubricating oil. If braking the engine to stop it or slow it by increasing the load does not work, find some means to stop the airflow to it.

To stop the airflow, activate engine shutdown devices (such as air intake flappers) to cover the air intake, or find some way of securing the air to the blower intake. If trying to secure the air to the blower intake, make certain that the covering will not be sucked into the blower, as this would cause an additional casualty.

NOTE: Do not attempt to use a portable carbon dioxide ($CO_2$) fire extinguisher to secure a diesel engine. The carbon dioxide ($CO_2$)

in the portable extinguisher will have little or no effect on the diesel engine. This is because the volume of air consumed by the diesel engine will be far greater than the volume of $CO_2$ contained in the extinguisher bottle.

## SUMMARY

In this chapter, we have discussed some basic operating procedures that may be able to be applied to the type of unit to which assigned. Our intent in this chapter was to provide general knowledge in regard to engine-room operations and to inform that the Navy Diesel Engineers are directed to the EOSS for specific applications. As recalled, the EOSS provides detailed operational and casualty control procedures for propulsion and auxiliary evolutions. From the information in this chapter, recognize the importance of the EOSS and from the information in this chapter and how it relates to normal operation and casualty control. Remember, casualty control that is performed properly will reduce equipment downtime and needless deterioration of engine components. When casualty control is NOT done properly, losses in terms of equipment, the mission, and even the operator's life or the life of a shipmate can result.

You should also be able to recognize the fundamental starting, operating, and stopping procedures used for a diesel engine under normal operating conditions and some of the emergency and casualty prevention procedures used under adverse circumstances.

# CHAPTER 3

# ENGINE PERFORMANCE AND EFFICIENCY

The prime concern of Navy Diesel Engineers is to keep the machinery for which he is responsible operating in the most efficient manner possible. From experience and training, they know that engine efficiency and performance depend upon much more than just operating the throttle and changing oil at prescribed intervals. The preceding chapters have covered many of the casualties, which may occur to reduce the power output of an engine. They have learned how to prevent the occurrence of many of these casualties. As they gained experience and understanding, they had to train other engineers. The engineers they trained frequently came up with the inevitable question, "Why?" They had to possess the ability to answer the many questions on why an engine does or does not perform efficiently.

The information given here is by no means all-inclusive, but it should serve as an aid to a better understanding of the factors that influence the efficiency and performance of an engine. The information in this chapter covers in more detail the factors influencing engine performance and efficiency, and concludes with a section on indicator cards.

## Engine Performance

In addition to the mechanical difficulties, engine performance is affected by other factors, some of which are inherent in engine design while the operator can control others. Note that the conditions for the two type engines are somewhat similar, except for some differences existing between factors dealing with fuel and ignition.

## Power Limitations

The design of an engine limits the amount of power that an engine can develop. Other limiting factors are the mean effective pressure, the length of stroke, the cylinder bore and the number of revolutions per minute (rpm) (piston speed) of the engine. The latter, piston speed, is limited by the frictional heat and by the inertia of the moving parts.

Diesel engineers learn how heat losses, efficiency of combustion, volumetric efficiency, and the proper mixing of fuel and air limit the power which a given engine cylinder can develop. They become familiar with the factors which cause overloading of an engine and unbalance between engine cylinders. They know the symptoms, causes, and effects of cylinder load unbalance and the steps that are necessary to maintain an equal load on each cylinder.

They know what is meant by engine efficiency and know how the various types of efficiencies and losses are used in analyzing the internal combustion process. They are also familiar with those factors that may cause the efficiencies to increase or decrease, and with the ways these variations affect engine performance.

## Mean Effective Pressure

The mean effective pressure (MEP) is the average pressure exerted on the piston during each power stroke, and is determined from a formula by means of a planimeter. There are two kinds of mean effective pressure: indicated mean effective pressure (IMEP), which is developed in the cylinder and can be measured; and brake mean effective pressure (BMEP), which is computed from the brake horsepower (bhp) delivered by the engine.

## Length of Stroke

The distance a piston travels between dead centers (top dead center, bottom dead center) is known as the length of stroke. This distance is one of the factors that determines the piston speed. In some modern diesel engines, piston speeds may reach about 1600 feet per minute.

## Cylinder Bore

Bore is used to identify the diameter of the cylinder. The diameter must be known in order to compute the area of the piston crown. It is upon this area that the pressure acts to create the driving force. This pressure is calculated and expressed for an area of one square inch. The ratio of length of stroke to cylinder bore is fixed in engine design; in almost all cases, the stroke is greater than the bore.

## Revolutions Per Minute

The crankshaft rotates at this speed. Since the piston is connected to the shaft, the revolutions per minute (rpm), along with the length of the stroke, determine piston speed. During each revolution, the piston completes one up-stroke and one down-stroke; therefore, piston speed is equal to rpm times twice the length of the stroke. This speed is usually expressed in feet per minute (fpm). If the stroke is 10.5 inches, and the speed of rotation is 720 rpm, the piston speed is computed as follows:

$$720 \text{ times } 2 \text{ times } \frac{10.5}{12} = 1260 \text{ fpm}$$

## Horsepower Computation

The power developed by an engine depends upon the type of engine as well as the engine's speed. Remember that a cylinder of a single-acting, 4-stroke cycle engine will produce one power stroke for every two crankshaft revolutions, while a single-acting, 2-stroke cycle engine produces one power stroke for each revolution.

**Indicated Horse Power**

The power developed within a cylinder can be calculated by measuring the imep and engine speed. (The rpm of the engine is converted to the number of power strokes per minute.) With the bore and stroke known (available in engine manufacturers' instruction manuals), the horsepower can be computed for the type engine involved. This power is called indicated horsepower (ihp) because it is obtained from the pressure measured with an engine indicator. Power loss due to friction is not considered in computing ihp.

Using the factors, which influence the engine's capacity to develop power, the general or standard formula for calculating ihp, is as follows:

$$IHP = \frac{P \times L \times A \times N}{33,000}$$

P=Mean indicated pressure, in psi
L=Length of stroke, in feet A=Effective area of the piston, in square inches
N=Number of power strokes per minute
   33,000-unit of power (one horsepower), or foot-pounds per minute

To illustrate the use of this formula, assume that a 12-cylinder, 2-stroke cycle, single-acting engine has a bore of 8.5 inches and a stroke of 10 inches. Its rated speed is 744 rpm. With the engine running at full load and speed, the imep is measured and found to be 105 psi. What is the indicated horsepower developed by the engine?

In this case,

$$P = 105; L = \frac{10}{12}; A = 3.1416 \left(\frac{8.5}{2}\right)^2; N = 744$$

Substituting these amounts in the formula, you have

$$\text{IHP} = \frac{105 \times \frac{10}{12}; \times 3.1416 \left(\frac{8.5}{2}\right)^2 \times 744}{33,000}$$

$$= 111.9$$

This amount represents the horsepower developed in only one cylinder; total horsepower for the engine will equal 12 times 111.9, or approximately 1343.

**Brake Horsepower (BHP)**. This power, sometimes called shaft horsepower, is the amount available for useful work. Brake horsepower is less than indicated horsepower because of the various power losses, which occur during engine operation. To obtain the brake or shaft horsepower delivered as useful work by an engine, the sum total of all mechanical losses must be deducted from the total indicated horsepower.

### Cylinder Performance Limitations

The factors, which limit the power that a given cylinder can develop, are piston speed and mean effective pressure. The piston speed, as stated before, is limited by the inertia forces set up by the moving parts and frictional heat; in the case of the mean effective pressure, the limiting factors are as follows:

1. Heat losses and efficiency of combustion

2. Volumetric efficiency, or the amount of air charged into the cylinder and the degree of scavenging

3. Mixing of the fuel and air

The manufacturer or BuShips prescribes the limiting mean effective pressures, both brake and indicated which should never be exceeded. In a direct-drive ship, the mean effective pressures developed are determined by the rpm of the power shaft. In electric-drive ships, the horsepower and bmep can be determined by a computation based on readings from electrical instruments, and generator efficiency.

## Cylinder Load Balance

In order to ensure a balanced, smooth-operating engine, the general mechanical condition of the engine must be properly maintained so that the power output of the individual cylinders is within the prescribed limits at all loads and speeds. In order to have a balanced load on the engine, each cylinder must produce its share of the total power developed. If the engine is developing its rated full power, or nearly so, and one cylinder or more is producing less than its share, obviously the remainder of the cylinders will become overloaded.

Using the rated speed and bhp, it is possible to determine for each INDIVIDUAL CYLINDER a rated bmep, which may not be exceeded without overloading the cylinder. If ENGINE rpm drops below the rated speed, then the cylinder bmep generally drops to a lower value. The bmep should never exceed the normal mep at lower engine speed. Usually, it should be somewhat lower if the engine speed is decreased.

Some engine manufacturers design the fuel systems so that it is impossible to exceed the rated bmep to any extent. This is done by installing a positive stop to limit the maximum throttle or fuel control. This regulates the maximum amount of fuel that can enter the cylinder and therefore the maximum load of the cylinder.

Engines used by the Navy are generally rated higher than those for industrial use, in order to meet emergencies. The economical speed for most of the Navy's diesel engines is approximately 90 percent of the rated speed. For this speed, the best load conditions have been found to be from 70 percent to 80 percent of the rated load or output. On this basis, if an engine is operated at an 80–90 combination (80 percent of rated load at 90 percent rated speed) the parts will give a longer life and the engine remain cleaner and in better operating condition.

Diesel engines do not operate well at exceedingly low bmep such as that occurring at idling speeds. Idling an engine tends to gum up parts associated with the combustion spaces. Operating an engine at idling speeds for long periods will result in the necessity for cleaning and overhauling much sooner than operation at 50 to 100 percent of load.

**Symptoms of Unbalance**

Evidence of an unbalanced condition existing between the cylinders of an engine may be indicated by the following symptoms:

1. Black exhaust smoke. When this occurs, it is not possible to determine immediately whether the entire engine or just one of the cylinders is overloaded. To determine which cylinder is overloaded, open the indicator cock on individual cylinders and check the color of the exhaust.

2. High exhaust temperatures. When the temperatures of exhaust gases from individual cylinders become higher than normal, it is an indication of an overload within the cylinder. When the temperature of the gases in the exhaust header becomes higher than usual, it indicates that all cylinders are probably overloaded. A frequent check on the pyrometer will indicate accurately whether each cylinder is firing properly and carrying its share of the load.

Any sudden change in the exhaust temperature of any cylinder should be investigated immediately. The difference in exhaust temperatures between any two cylinders should not exceed the limits prescribed in the engine manufacturer's instruction manual.

3. High lubricating oil and cooling water temperatures. If the temperature gages for these systems show an abnormal rise in the temperature, an overloaded condition may exist. The causes of an abnormal temperature in these systems should be determined and corrected immediately if engine efficiency is to be maintained.

4. Excessive heat. In general, such a condition in any part of the engine may indicate overloading. An overheated bearing may be the result of an overloaded cylinder; or an abnormally hot crankcase could result from overloading the engine as a whole.

5. Excessive vibration or unusual sound. If no cylinders are developing an equal amount of power, the forces exerted by individual pistons will be unequal. When this occurs, the unequal forces may cause an uneven turning moment to be exerted on the crankshaft, and vibrations will be set up. Through experience, you can learn to tell by the vibrations and sound of an engine when a poor distribution of load exists. Use every opportunity possible to observe engines running under all conditions of loading and performance.

**Causes of Unbalance**

An engine must be kept in excellent mechanical condition if unbalance is to be prevented. A leaky valve or fuel injector, leaky compression rings, or any other such mechanical difficulties will make it impossible for you to balance the load unless you secure the engine and dismantle at least a part of it. In other words, an engine must be placed in proper mechanical condition before the load can be balanced.

To obtain equal load distribution between individual cylinders, the clearances, tolerances, and the general condition of all parts that affect the cycle must be maintained so that very little, if any, variation exists between individual cylinders. In this connection, unbalance will occur unless the following are as nearly alike as possible for all cylinders:

1. Compression pressures.

2. Fuel injection timing.

3. Quantity and quality of fuel injected.

4. Firing pressures.

5. Valve timing and lift.

6. Indicator cards, when practicable to take them.

When unbalance occurs, correction usually involves repair, replacement, or adjustment of the affected part or system. Before any adjustments are made to eliminate unbalance, it must be determined beyond all doubt that the engine is in proper mechanical condition. When an engine is in good mechanical condition, few, if any, adjustments will be required. However, after an overhaul in which piston rings or cylinder liners have been renewed, considerable adjustment may be necessary. Until the rings become properly seated, some lubricating oil will leak past the rings into the combustion space. This excess oil will burn in the cylinder, giving an incorrect indication of fuel oil combustion. If the fuel pump is set for normal compression, and the rings have not seated properly, the engine will become overloaded.

As the compression rises to normal pressures, there will be an increase in the power developed, as well as in the pressure and temperature under which the combustion takes place. Therefore, when an overhaul has been completed, the engine instruments must be carefully watched until the rings are seated, and adjustments made as necessary. Frequent compression tests will serve as a helpful aid in making the necessary adjustments. Unless an engine is so equipped that compression can be readily varied, the engine should be operated under light load until you are sure that the rings are properly seated.

## Effect of Unbalance

From the preceding discussion, it can be readily seen that, in general, the result of unbalance will be overheating of the engine. The clearances established by the engine designer allow for sufficient expansion of the moving parts when the engine is operating at the designed temperatures, but an engine operating at temperatures in excess of those for which it was designed is subject to many casualties. Excessive expansion soon leads to seizure and burning of the engine parts. If the temperatures rise above the flash point of the lubricating oil vapors in the crankcase, an explosion may result. High temperature may destroy the oil film between adjacent parts, and the resulting increased friction will further increase the temperature.

Since power is directly proportional to the mean effective pressure developed in a cylinder, any increase in mean effective pressure will cause a corresponding increase in power. If the mean effective pressures in the individual cylinders vary, power will not be evenly distributed among the cylinders.

The quality of combustion obtained depends upon the heat content of the fuel, and the amount of heat available for power depends upon temperature. Temperature varies directly as pressure; therefore, a decrease in pressure will result in a decrease in temperature, and in poor combustion. The results of poor combustion will be lowered thermal efficiency and reduced engine output.

Cylinder load balance is essential if the desired efficiency and performance of an engine is to be obtained. To avoid the harmful effects of overloading and unbalancing of load, the load on an engine should be properly distributed among the working cylinders; and no cylinder, or the engine itself, should ever be overloaded.

In general, load balance in engine can be maintained if the following procedure is observed:

1. Maintain the engine in proper mechanical condition.

2. Adjust the fuel system according to the manufacturer's instructions.

3. Operate the engine within the temperature limits specified in appropriate instructions.

4. Keep cylinder temperatures and pressures as evenly distributed as possible.

5. Train yourself to recognize the symptoms of serious engine conditions.

**Engine Efficiency**

Engine efficiency is the amount of power developed compared to the energy input, which is measured by the heating value of the fuel consumed. In other words, the term "efficiency" is used to designate the relationship between the result obtained and the effort expended to produce the result. The term "compression ratio" is frequently used in connection with engine performance and the various types of efficiencies. From the study of the principles of internal combustion, recall that compression ratio is the ratio of the volume of air above the piston when it is at the bottom dead center position to the volume of air above the piston when it is at the top dead center position.

**Efficiencies**

The principle efficiencies considered in the internal combustion process are cycle, thermal, mechanical, and volumetric.

**Cycle Efficiency**

The efficiency of any cycle is equal to the output divided by the input. The efficiency of the diesel cycle is considerably higher than the Otto or constant volume cycle because of the higher compression ratio and because combustion starts at a higher temperature. In other words, the heat input is at a higher average temperature. Theoretically, the gasoline engine using the Otto cycle would be more efficient than the diesel engine if equivalent compression ratios could be used. However, engines operating on the Otto cycle cannot use a compression ratio comparable to those of diesel engines because the fuel and air are drawn into the cylinder together and compressed. If comparable compression ratios were used, the fuel would fire or detonate before the piston reached the correct firing position.

Since temperature and amount of heat content, which is available for power, are proportional to each other, the cycle efficiency is actually computed from measurements made of the temperature. The specific heat of the mixture in the cylinder is either known or assumed, and when combined with the temperature, the heat can be calculated at any instant.

**Thermal Efficiency**

This may be regarded as a measure of the efficiency and completeness of combustion of the fuel, or, more specifically, the ratio of the output or work done by the working substance in the cylinder in a given time to the input or heat energy of the fuel supplied during the same time. Generally, two kinds of thermal efficiency are considered for an engine: indicated thermal efficiency and overall thermal efficiency.

Since the work done by the gases in the cylinder is called indicated work, the thermal efficiency determined by its use is often called *indicated thermal efficiency.* If all the potential heat in the fuel could be delivered as work, the thermal efficiency would be 100 percent. Because of the various losses, this percent is not possible in actual installations.

If the amount of fuel injected is known, the total heat content of the injected fuel can be determined from the heating value or Btu per pound, of the fuel; the thermal efficiencies for the engine can then be calculated. From the mechanical equivalent of heat, (778 foot-pounds equal 1 Btu and 2545 Btu equal 1 hp-hr), the number of foot-pounds of work contained in the fuel can be computed. If the amount of fuel injected is measured over a period, the rate at which the heat is put into the engine can be converted into potential power. Then, if the indicated horsepower developed by the engine is calculated as previously explained, the INDICATED THERMAL EFFICIENCY (ITE) can be computed by the following expression:

Indicated thermal efficiency =

Indicated hp x2545 Btu per hour per hp
Rate of heat input of fuel in Btu per hour x 100

For example, assume that the same engine used as an example in computing indicated horsepower consumes 360 lb (approximately 50 gallons) of fuel per hour, and the fuel has a value of 19,200 Btu per pound. What is the indicated thermal efficiency of the engine?

The work done per hour when 1343 ihp are developed is 1343 x 2545 or 3,417,935 Btu. The heat input for the same time is 360 x

19,200, or 6,912,000 Btu. Then, by the above expression, the indicated thermal efficiency is as follows:

Indicated thermal efficiency =

$$\frac{1343 \times 2545}{360 \times 19,200} \times 100, \text{ or } \frac{3,417,935}{6,912,000} \times 100, \text{ or } 49.4\%$$

The other type of thermal efficiency—over ALL THERMAL EFFICIENCY considered for an engine is a ratio similar to indicated thermal efficiency, except that the useful or shaft work (brake horsepower) is used. Therefore, over-all efficiency (often called brake thermal efficiency) is computed by the following expression:

Overall thermal efficiency =

$$\frac{\text{Brake horsepower}}{\text{Heat input of fuel}} \times 100$$

Converting these factors into the same units (Btu), the expression is written as power output in Btu divided by fuel input in Btu.

For example, if the engine used in the preceding problem delivers 900 brake horsepower (determined by the manufacturer) what is the over-all thermal efficiency of the engine?

$$1 \text{ hp-hr} = 2545 \text{ Btu}$$

$$900 \text{ bhp} \times 2545 \text{ Btu per hp-hr} = 2,290,500 \text{ Btu output per hour.}$$

Substituting factors already known, over-all thermal efficiency is computed as follows:

Over-all thermal efficiency=

$$\frac{2,290,500}{6,912,000}=0.331, \text{ or } 33.1\%$$

Compression ratio influences the thermal efficiency of an engine. Theoretically, the thermal efficiency increases as the compression ratio is increased. The maximum value of a Diesel engine compression ratio is determined by the compression required for starting; and this compression is, to a large extent, dependent on the type of fuel used. The maximum value of the compression ratio is not limited by the fuel used, but is limited by the strength of the engine parts and the allowable engine weight per brake horsepower output.

**Mechanical Efficiency**

This is the rating that shows how much of the power developed by the expansion of the gases in the cylinder is actually delivered as useful power The factor which has the greatest effect on mechanical efficiency is friction within the engine. The friction between moving parts in an engine remains practically constant throughout the engine's speed range. Therefore, the mechanical efficiency of an engine will be highest when the engine is running at the speed at which maximum brake horsepower is developed. Since power, output is brake horsepower, and the maximum horsepower available is indicated horsepower, then,

$$\text{Mechanical efficiency} = \frac{\text{brake horsepower}}{\text{indicated horsepower}} \times 100$$

During the transmission of indicated horsepower through the piston and connecting rod to the crankshaft, the mechanical losses that occur may be due to friction, or they may be due to power absorbed. Friction losses occur because of friction in the various bearings, or between piston and piston rings and the cylinder walls. Power is absorbed by valve and injection mechanisms, and by various auxiliaries, such as the lubricating oil and water circulating pumps and the scavenge and supercharge blowers. As a result, the power delivered to the crankshaft and available for doing useful work (bhp) is less than indicated power.

The mechanical losses that affect the efficiency of an engine may be called frictional horsepower (fhp) or the difference between ihp and bhp. The flip of the engine used in the preceding examples, then, would be 1343 (ihp)–900 (bhp) 443 fmp, or 33 percent of the ihp developed in the cylinders. Then, using the expression for mechanical efficiency, the percentage of power available at the shaft is computed as follows:

$$\textbf{Mechanical efficiency} = \frac{900}{1343} = .67 \times 100 = 67\%$$

When an engine is operating under part load, it has a lower mechanical efficiency than when operating at full load. The explanation for this is that most mechanical losses are almost independent of the load, and therefore, when load decreases, ihp decreases relatively less than bhp. Mechanical efficiency becomes zero when an engine operates at no load because then bhp = 0, but ihp is not zero. In fact, if bhp is zero and the expression for flip is used, ihp is equal to fhp.

To show how mechanical efficiency is lower at part load, assume the engine used in preceding examples is operating at 3/4-load. Brake horsepower at 3/4-load is 900 x 0.75 or 675. Assuming that flip does not change with load, fhp = 443. The ihp is, by expression, the sum of bhp and fhp.

$$\textbf{ihp} = 675 \text{ (bhp)} + 443 \text{ (fhp)} = 1118$$

Mechanical efficiency = 675 + 1118 =0.60, or 60 percent; this is appreciably lower than the 67 percent indicated for the engine at full load.

Brake mean effective pressure (bmep), is a useful concept when dealing with mechanical efficiency. Bmep can be obtained if the standard expression for computing horsepower (ihp) is applied to bhp instead of ihp and the mean pressure (p) is designated as bmep.

$$bhp = \frac{(bmep) \times L \times A \times N}{33,000}$$

or

$$bmep = \frac{33,000 \times bhp}{L \times A \times N}$$

From the relations between brmep, bhp, ihp, and mechanical efficiency, by designating indicated mean effective pressure by imep in the expression, one can also show:

Bmep = imep x mechanical efficiency

To illustrate this, the bmep for the engine in preceding examples at full load and ¾ load is computed as follows:

$$bmep = \frac{33,000 \times \frac{(bhp)}{(12)}}{L \times A \times N} = \frac{33,000 \times \frac{(900)}{(12)}}{\frac{10}{12} \times 56.74 \times 744} = 70 \text{ psi}$$

or

$$Bmep = imep \times mechanical\ efficiency$$
$$= 105 \times 67, \text{ or } 70 \text{ psi}$$

Bmep gives an indication of the load an engine carries, and what the output is for piston displacement. As the bmep for an engine increases, the engine develops greater horse-power per pound of weight. For a given engine, bmep changes in direct proportion with the load.

## Volumetric Efficiency

The volumetric efficiency of a 4-stroke engine is the relationship between the quantity of intake air and the piston displacement. In other words, volumetric efficiency is the ratio between the charge that actually enters the cylinder and the amount that could enter under ideal conditions. Piston displacement is used since it is difficult to measure the amount of charge that would enter the cylinder under ideal conditions.

An engine would have 100 percent volumetric efficiency if, at atmospheric pressure and normal temperature, an amount of air exactly equal to piston displacement could be drawn into the cylinder. This is not possible, except by super- charging, because the passages through which the air must flow offer a resistance, the force pushing the air into the cylinder is only atmospheric, and the air absorbs heat during the process. Therefore, volumetric efficiency is determined by measuring (with an orifice or venturi type meter) the amount of air taken in by the engine, converting the amount to volume, and comparing this volume to the piston displacement.

Volumetric efficiency =

$$\frac{\text{Volume of air admitted to cylinder}}{\text{Volume of air equal to piston displacement}} \times 100$$

The concept of volumetric efficiency does not apply to 2-stroke cycle engines. Instead, the term "scavenge efficiency" is used, which shows how thoroughly the burned gases are removed and the cylinder filled with fresh air. As in the case of a 4-stroke cycle engine, it is desirable that the air supply be sufficiently cool. Scavenge efficiency depends largely upon the arrangement of the exhaust; scavenge air ports, and valves.

## Engine Losses

As the heat content of a fuel is transformed into useful work, during the combustion process, many different losses take place. These losses can be divided into two general classifications: thermodynamic and mechanical. The net useful work delivered by an engine is the result obtained by deducting the total losses from the heat energy input.

*Thermodynamic Losses.* Losses of this nature are a result of the following: loss to the cooling and lubricating systems; loss to the surrounding air; loss to the exhaust; and loss due to lack of perfect combustion.

Heat energy losses from both the cooling water systems and the lubricating oil system are always present. Some heat is conducted through the engine parts and radiated to the atmosphere or picked up by the surrounding air by convection. The effect of these losses varies according to the part of the cycle in which they occur. The heat of the jacket cooling water cannot be taken as a true measure of heat losses, since all this heat is not absorbed by the water. Some heat is lost to the jackets during the compression, combustion, and expansion phases of the cycle; some is lost (to the atmosphere) during the exhaust stroke; and some is absorbed by the walls of the exhaust passages.

Heat losses to the atmosphere through the exhaust are unavoidable. This is because the engine cylinder must be cleared of the hot exhaust gases before the next air intake charge can be made. The heat lost to the exhaust is deter- mined by the temperature within the cylinder when exhaust begins. The amount of fuel injected and the weight of air compressed within the cylinder are controlling factors. Improper timing of the exhaust valves, whether early or late, will result in increased heat losses. If early, the valve releases the pressure in the cylinder before all the available work is obtained; if late, the necessary amount of air for complete combustion of the next charge cannot be realized, although a small amount of additional work may be obtained. Proper timing and seating of the valves is essential in order to maintain heat loss to the exhaust at a minimum.

Heat losses due to imperfect or incomplete combustion have a serious effect on the power that can be developed in the cylinder. Because of the short interval of time necessary for the cycle in modern engines, complete combustion is not possible; but heat losses can be kept to a minimum if the engine is kept in proper adjustment. It is often possible to detect incomplete combustion by watching for abnormal exhaust temperatures and changes in the exhaust color, and by being alert for unusual noises in the engine.

**Mechanical Losses**

There are several kinds of mechanical losses, but all are not present in every engine. The mechanical or friction losses of an engine include bearing friction; piston and piston ring friction; pumping losses caused by operation of water pumps, lubricating pumps, and scavenging air blowers; power required to operate valves; etc. Friction losses cannot be eliminated, but they can be kept to a minimum by maintaining the engine in its best mechanical condition. Bearings, pistons, and piston rings should be properly installed and fitted, shafts must be in alignment, and lubricating and cooling systems should be at their highest operating efficiency.

Remember that the total of these mechanical losses must be deducted from ihp of the engine in order to determine actual bhp.

## Summary

To understand the various factors that influence engine performance and efficiency, a thorough knowledge of the internal combustion process is necessary. Once the combustion process is understood, it will be much easier to appreciate the parts played by such factors as engine design, engine operating conditions, fuel characteristics, fuel injection, ignition, pressures and temperatures, and compression ratios. This chapter provides some of the information necessary for a better understanding of how engine performance and efficiency is affected by many factors. The Navy Diesel Engineer is able to gain complete understanding of such factors only through continued study and practical experience.

They know how the power, which an engine can develop, is limited by mean effective pressure, length of piston stroke, cylinder bore, and engine speed, and how these factors are used in determining the power developed by an engine. Learn how heat losses, efficiency of combustion, volumetric efficiency, and the proper mixing of fuel and air limit the power which a given engine cylinder can develop. They are familiar with the factors which cause overloading of an engine and unbalance between engine cylinders. They know the symptoms, causes, and effects of cylinder load unbalance and know what is necessary if an equal load is to be maintained on each cylinder.

Navy Diesel Engineers know what is meant by engine efficiency and know how the various types of efficiencies and losses are used in analyzing the internal combustion process. They are familiar with the factors that may cause the various efficiencies to increase or decrease, and the way in which these variations affect engine performance.

Engine indicators and the diagrams they produce are of extreme importance when analyzing what takes place in the cylinder of an engine equipped with indicator gear. They know the principles of operation of an indicator, and know the relationship of each part of a diagram with the events which take place in a cycle. You should be able to use the information provided by an indicator diagram for computing the power developed by an engine. Be able to analyze an indicator diagram and determine what changes, both pressure and volume, take place within an engine cylinder. This will serve as an aid in determining if the various processes are taking place in the cylinder in the proper manner. When a diagram shows any deviation from the standard diagram for an engine, this is an indication that adjustments or repairs are necessary.

Keep in mind that rules must be established for the analysis and interpretation of the indicator diagrams for a given engine, and that such rules should be based on a diagram representing the normal operation of the engine. Know how the information available from an indicator card can be used as an aid in keeping an engine performing efficiently.

# CHAPTER 4

# ENGINE TROUBLESHOOTING

In troubleshooting an internal combustion engine, whether diesel or gasoline, the procedures are somewhat similar. In many instances, the information, which follows, will apply to both types of engine. However, principal differences, which do exist, will be described.

This chapter is concerned with troubles encountered in starting an engine. The troubles listed here are chiefly of the kind that can be corrected without major overhaul or repair. The troubles discussed in the following chapter are those that can be identified by erratic operation of the engine, by warnings by the instruments, or by inspection of the engine parts and systems. There is also a section devoted to those systems of the gasoline engine, which are different from those of the diesel engine. Keep in mind that the troubles listed here are general a may or may not apply to a particular diesel engine. When working with a specific engine, check the manufacturer's instruction book.

## Qualifications of the Trouble Shooter

Complete failure of a power plant at a crucial moment may imperil both ship and crew. Minor engine trouble, if not recognized and corrected as soon as possible, may develop into a major breakdown. Therefore, it is essential that every operator of an internal combustion engine train himself to be a successful troubleshooter.

It may happen that an engine will continue to operate even when a serious casualty is imminent. However, if troubles are impending, there will probably be symptoms present, and the success of a troubleshooter depends partially upon his ability to recognize these symptoms when they occur. The good operator uses most of his senses to detect trouble symptoms. He may see, hear, smell, or feel the symptoms, which serve as a warning of trouble to come. Of course, common sense is also a requisite. Another factor upon which the success of a troubleshooter depends is his ability to locate the trouble after once deciding something is wrong with the equipment. Then he must be able to determine as rapidly as possible what corrective action must be taken. In learning to recognize and locate engine troubles, experience is the best teacher.

Instruments play an important part in the detection of engine troubles. The engine operator should read the instruments and record their indications regularly. If the recorded indications vary radically from those specified by engine operating instructions, it is a warning that the engine is not operating properly and that some type of corrective action must be taken. Familiarity with the specifications given in engine operating instructions is essential, especially those pertaining to temperatures, pressures, and speeds. When instrument indications vary considerably from the specified values, the operator should know the probable effect on the engine. When variations occur in instrument indications, be sure before taking any corrective action that such variations are not the fault of the instrument.

Periodic inspections are also essential to detect engine troubles. Failure of visible parts, presence of smoke, or leakage of oil, fuel, or water can be discovered by such inspections. Cleanliness is probably one of the greatest aids to the detection of leakage. When engine casualties occur suddenly while a ship is under way, the Lead Diesel Engineer must make decisions quickly and accurately. He must decide immediately whether to secure the engine (with proper permission) or to let it operate. An incorrect decision to let it operate might result in extensive damage. The operator must be an expert in determining the seriousness of the trouble because his decision to secure or to continue operation is based on the severity of the trouble and the immediate need for power by the vessel.

When an engine is secured because of a trouble, the procedure for repairing the casualty follows an established pattern, if you know where the trouble exists. If diesel engineers do not know the location of the trouble, then they must begin a search for the cause. To inspect every part of an engine whenever a trouble occurs would be an endless task. The cause of a trouble can be found much more quickly if a systematic and logical method of inspection is followed. A well-trained troubleshooter can isolate a trouble by identifying it with one of the engine systems. Once the trouble has been associated with a particular system, the next step is to trace out the system until they find the cause of the trouble. Troubles generally originate in only one system, but remember that troubles in one system may cause damage in another system or to component engine parts. When a casualty involves more than one system of the engine, trace each system separately and make corrections as necessary. It is obvious that diesel engineers must know the construction, function, and operation of the various systems as well as the parts of each system for a specific engine before you can satisfactorily locate and remedy troubles.

Diesel engineers continually strive to keep the equipment for which they are responsible operating satisfactorily. Troubles frequently prevent such operation. Even though there are many troubles which may affect the operation of a Diesel engine, satisfactory performance depends primarily on the presence of sufficiently high compression pressure and the injection of the right amount of fuel at the proper time. Proper compression depends on the pistons, piston rings, and valve gear, while the right amount of fuel obviously depends on the fuel injectors and actuating mechanism. Such troubles as lack of engine power, unusual or erratic operation, and excessive vibration may be caused by either insufficient compression or faulty injector action.

Diesels engineers have learned a lot about engine troubles from experience as well as study while working their way up to a higher rate. They know the symptoms and causes of engine casualties and how to correct them. The following information may serve as a review and as a general guide when you instruct men in a lower rate on how to keep equipment operating properly.

Many of the troubles encountered by an engine operator can be avoided if the prescribed instructions for starting and operating an engine are followed. The list of troubles, which follow, cannot be considered complete and they do not necessarily apply to all diesel engines because of differences in design. Specific information on trouble shooting all the diesel engines used by the Navy would require more space than available here.

Even though a successful troubleshooter generally associates a trouble with a particular system or assembly, the troubles are discussed according to either when they might be encountered, before or after the engine starts. The troubles are indicative of the system to which they apply and require no further identification.

**Engine Fails to Start**

The troubles, which prevent an engine from starting, may be grouped under the following heads: (1) the engine can neither be cranked nor barred over, (2) the engine cannot be cranked but it can be barred over, and (3) the engine can be cranked, but it still fails to start.

## Troubles that may prevent a Diesel from Starting

| ENGINE WILL NOT START | | |
|---|---|---|
| ENGINE CANNOT BE CRANKED NOR BARRED OVER | ENGINE CANNOT BE CRANKED BUT CAN BE BARRED OVER | ENGINE CAN BE CRANKED BUT FAILS TO START |
| Improperly engaged jacking gear | Depleted air supply | Improper throttle setting |
| Seized piston | Closed airline valve | Contaminated fuel |
| Obstructions in cylinder | Engaged locking gear interlock | Insufficient fuel supply |
| Improper bearing fit | Faulty air starting distributor | Improper fuel |
| | Faulty cylinder air- starting valves | Improper fuel system timing |
| | | Insufficient compression |
| | | Tripped overspeed device |
| | | Inoperative governor |
| | | Inoperative cold start |
| | | Insufficient cranking speed |

## Engine Cannot Be Cranked or Barred Over

Most pre-starting instructions specify that the crankshaft of an engine should be turned one or more revolutions before starting power is applied. If the crankshaft cannot be turned over, check the turning or jacking gear to make sure that it is properly engaged. If the jacking gear is properly engaged, and the crankshaft still fails to turn over, check to see if the cylinder test (relief) valves or indicator valves are closed and holding water or oil in the cylinder.

## Seizing

When turning gear operates properly, and the cylinder test valves are open, but the engine cannot be cranked or barred over, the source of the trouble will probably be of a much more serious nature. A piston or other part may be seized, or a bearing may be fitting too tightly. Sometimes the difficulty cannot be remedied except by removing a part or an assembly.

## Defective Piston

Some engines have ports through which pistons can be inspected. If inspection reveals that the piston is defective, the assembly must be removed. If the condition of an engine without cylinder ports indicates that a piston inspection is required, the whole assembly must be taken out of the cylinder.

Engine bearings have to be carefully fitted according to the manufacturer's instructions. When an engine cannot be jacked over because of an improperly fitted bearing, this means that someone failed to follow instructions when the unit was being reassembled. It also means that proper supervision was present during reassembly. The partial or perhaps complete overhaul required to correct the bearing fit not only involves extra time and work, but it may also be a reflection upon the dependability of you, as a supervisor.

## Engine Cannot Be Cranked but Can Be Barred Over

Most of the troubles which prevent the cranking of an engine but are not serious enough to prevent barring over can be traced to the air starting system, although other factors may prevent an engine from cranking. However, only troubles related to an air starting system are given in this chapter.

If an engine fails to crank when starting power is applied, first check the turning or jacking gear, to make sure that it is disengaged. If this gear is not the source of trouble, then the trouble is probably with the air starting system.

Although the design of different *air starting systems* varies, the function remains the same. In general, such systems must have a source of air such as the compressor or the ship's air system; a storage tank; air flask(s); an air timing mechanism; and a valve in the engine cylinder to admit the air during starting and to seal the cylinder while the engine is running.

**Defect in Timing Mechanism**

Air starting systems have a unit designed to admit starting air to the proper cylinder at the proper time. The name given to this unit, as well as the type of unit, may vary from one system to another. Such names as timer, distributor, air starting pilot valve, air starting distributor, and air distributor are applicable to this unit. The types of air timing mechanisms, which may be encountered, are the *direct mechanical lift*, the *rotary distributor*, and the *plunger type distributor valve*. The timing mechanism of an air starting system is relatively trouble free except as noted in the following cases.

The operation of the direct mechanical lift air timing mechanism involves the use of cams, push rods, and rocker arms, and the mechanism is subject to part failures similar to those occurring in corresponding major engine parts. Therefore, the causes of trouble in the actuating gear and the necessary maintenance procedures will be found under information covering similar parts of the major engine systems.

Most troubles are a result of improper adjustment. Generally, this involves the lift of the starting air cam or the timing of the air-starting valve. The starting air cam must lift the air-starting valve sufficiently so that when the engine is running; there will be proper clearance between the cam and cam valve follower. If proper clearance does not exist between these two parts, hot gases will flow between the valve and the valve seat, causing excessive heating of the parts. Since the starting air cam regulates the opening of the air-starting valve, those with adjustable cam lobes should be checked frequently to ensure that the adjusting screws are tight.

The proper values for lift, tappet clearance, and time of valve opening for a direct mechanical lift timing mechanism should be obtained from the instruction manual for the particular engine. Make adjustments only as specified.

The rotary distributor timing mechanism requires a minimum of maintenance, but there may be times when the unit will become inoperative, and will have to be disassembled and inspected. In such cases, a scored rotor, a broken spring, or improper timing generally causes the difficulty.

Since foreign particles in the air can cause scoring of the rotor, with excessive air leakage resulting, the air supply must be kept as clean as possible. Another cause of scoring is lack of lubrication. If the rotor in a hand-oiled system becomes scored because of insufficient lubrication, it may be that the equipment is at fault, or that lubrication instructions have not been followed. In either a hand-oiled or a pressure-lubricated system, check the piping and the passages, to see that they are open. When scoring is not too serious, the rotor and body should be lapped together. A thin coat of prussian blue can be used to determine whether the rotor contacts the distributor body.

A broken spring may be the cause of an inoperative timing mechanism if a coil spring is used to maintain the rotor seal. In such a case, replacing the spring is the only way to ensure an effective seal.

An improperly timed rotary distributor will keep an engine from cranking. Timing should be checked against information given in the specific engine instruction manual.

In a plunger type distributor valve timing mechanism, the valve requires little attention; however, it may stick occasionally, and prevent the proper functioning of the air starting system. On some engine installations, the pilot air valve of the distributor will not open, while on other installations this valve will not close. Dirt and gum deposits, broken return springs, or the lack of lubrication may cause the trouble. Deposits and lack of lubrication will cause the unit valve plungers to bind and stick in the guides, while a broken valve return spring prevents the plunger from following the cam profile. A distributor valve, which sticks, should be disassembled and thoroughly cleaned, and any broken springs must be replaced.

**Faulty Air Starting Valve**

The function of air starting valves is to admit starting air into the engine cylinder and then seal the cylinder while the engine is running. These valves may be of the pressure actuated or the mechanical lift type.

The valve may stick open for a number of reasons. A gummy or resinous deposit may cause the upper and lower pistons to stick in the cylinders. This deposit is formed by the oil and condensate, which may be carried into the actuating and the lower cylinders. Oil is necessary in the cylinders to provide lubrication and to act, as a seal but moisture should be eliminated. The formation of this resinous deposit can be prevented if the system storage tanks and water traps are drained as specified in operating instructions.

The deposit on the lower piston may be greater than in the actuating cylinder because of the heat and combustion gases that add to the formation if the valve remains open. When the upper piston is the source of trouble, it can usually be relieved, without removing the valve, by using light oil or Diesel fuel, and working the valve up and down. When this method is used to relieve a sticking valve, be sure that the valve surfaces are not burned or deformed. If this method does not relieve the sticking condition, the valve will have to be removed, disassembled, and cleaned. The type of valve shown requires removal of the cylinder head to disassemble the valve. Some types of pressure-actuated valves may be removed without cylinder head removal.

Pressure-actuated starting valves sometimes fail to operate because of broken or weak valve return springs. Replacement is generally the only solution to this condition; however, some valves are constructed with a means of adjusting spring tension. In valves so constructed, increasing the spring tension may eliminate the trouble.

Occasionally the actuating pressure of a valve will not release, and the valve will stick open or be sluggish in closing. The cause is usually clogged or restricted air passages. Combustion gases will enter the air passageways, burning the valve surfaces, and these burned surfaces usually have to be reconditioned before they will maintain a tight seal. Keeping the air passages open will eliminate extra maintenance work on the valve surfaces.

The mechanical lift type of air starting valve is subject to leakage, which, in general, is caused by the valve sticking open. Any air-starting valve that sticks or leaks creates a condition, which makes an engine hard to start. If the leakage in the air-starting valve is excessive, the resulting loss in pressure may be sufficient to prevent starting.

An overtightened packing nut can cause leakage in this type of valve. Overtightening the packing nut is sometimes employed to stop minor leaks around the valve stem when starting pressure is applied, but it may keep the air valve from seating. As in the pressure-actuated valve, return spring tension may be insufficient to return the valve to the valve seat after admitting the air change. If this occurs, gases from the cylinder will leak into the valve while the engine is running.

Obstructions such as particles of carbon between the valve and valve seat will hold the valve open, permitting combustion gases to pass.

A valve stem bent by careless handling during installation may prevent a valve from closing properly. If a valve hangs open for any of these reasons, hot combustion gases will leak past the valve and valve seat. This burns the valve and seat and may result in a leak between these two surfaces even though the original causes of the sticking are eliminated.

A leaking valve should be completely disassembled and inspected. It is subject to a resinous deposit similar to that found in a pressure-actuated air valve. A specified cleaning compound should be used for the removal of the deposit. Be sure the valve stem is not bent. Check the valve and valve seat surfaces carefully. Scoring or discoloration should be eliminated by lapping with a fine lapping compound. Jewelers' rouge or talcum powder with fuel oil may be used for lapping.

From the preceding description, it can readily be seen that the air starting system may be the source of many troubles that will prevent cranking an engine even though it can be barred over. A few of the troubles can be avoided if pre-starting and starting instructions are followed. One such instruction, sometimes overlooked, is that of opening the valve in the airline. Obviously, with this valve closed, the engine will not crank. See that the men learn to follow the instructions for starting an engine. If the engine will not crank, recheck the instructions for such an oversight as a closed valve, an empty air storage receiver, or an engaged jacking gear before starting any disassembly.

## Engine Cranks but Fails to Start

Even when the starting equipment is in an operating condition, an engine may fail to start. A majority of the possible troubles, which prevent an engine from starting, are associated with fuel and the fuel system. However, parts or assemblies, which are defective or inoperative, may be the Source of some trouble. Failure to follow instructions may be the cause of an engine failing to start. The corrective action is obvious for such items as leaving the fuel throttle in the OFF position and leaving the cylinder indicator valves open. As a supervisor, you cannot overstress the importance of following prescribed starting instructions and rechecking the procedure, if an engine fails to start.

## Foreign Matter in the Fuel Oil System

In the operation of an internal combustion engine, cleanliness is of paramount importance. This is especially true in the handling and care of diesel fuel oil. Impurities are the prime source of fuel pump and injection system trouble. Sediment and water cause wear, gumming, corrosion, and rust in a fuel system. Even though fuel oil is generally delivered, clean from the refinery, handling and transferring increases the chance of fuel oil becoming contaminated. The necessity of periodic inspection, cleaning, and care of fuel oil systems and filtering equipment should be continually emphasized.

Failure of an engine to start may be caused by air, dirt, or water in the fuel. Experience has already taught you much about the destructive effect of water on the parts of an engine. Corrosion frequently leads to replacement or at least to repair of the part. Steps should be taken to continually prevent the accumulation of water in a fuel system, not only to eliminate the cause of corrosion but also to ensure proper combustion in the cylinders. All fuel should be centrifuged, and fuel filter cases drained periodically to prevent excessive collection of water.

In general, fuel oil may contain water as a result of receiving contaminated fuel from a tanker, failure to strip fuel oil tanks properly after ballasting, leaking fuel oil storage tanks (either from the open sea or from an adjacent ballasted tank), following improper procedures when making routine tests for water in fuel oil tanks, or failure to properly centrifuge the fuel oil.

Water in fuel is injurious to the entire fuel system, and will cause irreparable damage in a short time. It not only corrodes the fuel injection pump, where close clearances must be maintained, but also corrodes and erodes the injection nozzles. The slightest corrosion will cause a fuel injection pump to bind and seize, and if not corrected will lead to excessive leakage. Water will cause the orifices of injection nozzles to erode until they will not spray the fuel properly, thus preventing proper atomization. When this occurs, incomplete combustion and engine knock results.

Air in the fuel system is another possible trouble, which may prevent an engine from starting. Even if starting is possible, air in the fuel, system will cause the engine to miss and knock, and perhaps stall.

When an engine fails to operate, stalls, misfires, or knocks, there may be air in the high-pressure pumps and lines. In many systems, the expansion and compression of such air may take place without the injection valves opening. If this occurs, the pump is air-bound. It can be determine if air exists in a fuel system by bleeding a small amount of fuel from the top of the fuel filter; if the fuel appears quite cloudy, it is likely that there are small bubbles of air in the fuel.

In troubleshooting the fuel system, remember that if air is entering a fuel line, the pressure within the fuel line must be lower than atmospheric pressure. The smallest of holes in the transfer pump suction piping will permit air to flow into the system in quantities sufficient to air bind the high-pressure pumps. Carefully check all fittings in the suction piping. A loose fitting or a damaged thread condition will allow air to enter the system. On installations where flanged connections are used, the condition of the gaskets should be checked. Tubing, especially copper, should be checked carefully for cracks that may result from constant vibration.

If an engine runs out of fuel, trouble can be expected from air that enters the fuel system. If a considerable quantity of air exists in the filter, a quick method of purging the system of air is to remove the filling plugs on top of the filter and pour in clean fuel oil until all air is displaced. Any air remaining in the system can then be removed by using the hand-priming pump.

Cranking the engine for a period not exceeding the specified cranking period will remove any small amount of air in the system; but if the engine does not start during this interval, cranking it further will only reduce or deplete the starting air supply.

On most installations, the hand-operated transfer pump may be used to remove air from the fuel system. Generally, the procedure is to remove air progressively from all parts of the system, starting with the suction line of the transfer pump and proceeding to the injection valves. However, the procedure varies slightly in different systems, depending on construction.

In the fuel system between the pump and strainers, break open the system. The pump is operated until all air is removed and only clear fuel flows from the line. Then the line is closed and the same procedure is repeated at such points in the system as between the strainers and the filters, between the filters and the high-pressure pumps, and at the overflow line connection on the high-pressure pump housing.

In small high-speed diesel engines, priming at the overflow connection may be all that is necessary. Since priming high-pressure lines is time consuming, attempt to start the engine before purging these lines. However, the engine must not be cranked for more than the specified interval of time. If the engine still fails to start, priming the high-pressure lines will be necessary.

*Insufficient Fuel Supply* may be the reason for an engine not starting. This condition may result from any one of a number of defective or inoperative parts in the system. Such items as a closed inlet valve in the fuel piping or an empty supply tank are more apt to be the fault of the operator than of the equipment. However, an empty tank may be caused by leakage in the lines or in the tank.

Leakage can usually be traced leakage in the low-pressure lines of a fuel system to cracks in the piping. Usually these cracks occur on threaded pipe joints at the root of the threads. Such breakage is caused by the inability of the nipples and pipe joints to withstand shock, vibration, and strains resulting from the relative motion between smaller pipes and equipment to which they are attached.

Metal fatigue can also be a cause of breakage; each vessel should have a systematic inspection of the installation of fittings and piping to determine if all parts are satisfactorily supported and sufficiently strong. In some instances, nipples may be connected to relatively heavy parts, such as valves and strainers, which are free to vibrate. Since vibration contributes materially to the fatigue of nipples, rigid bracing should be installed. When practicable, bracing should be secured to the unit itself, instead of to the hull or other equipment. Leakage in the high-pressure lines of a fuel system also results from breakage. This breakage usually occurs on either of the two end fittings of a line, and is caused by lack of proper supports or by excessive nozzle opening pressure.

Supports are usually supplied with an engine, and should not be discarded. Excessive opening pressure of a nozzle generally due to improper spring adjustment or to clogged nozzle orifices may rupture the high-pressure fuel lines. A faulty nozzle generally requires removal, inspection, and repair plus the use of a nozzle tester.

Leakage from fuel lines may be due also to improper re-placements or repairs. Spare high-pressure fuel lines should be kept on hand at all times. When a replacement is necessary, always use a line of the same length and diameter as the one removed. Varying the length and diameter of a high-pressure fuel line will change the injection characteristics.

In an emergency, high-pressure fuel lines can usually be satisfactorily repaired by silver soldering a new fitting to the line. Be sure the centers of the fitting and the line are properly located. Take extreme care to prevent the solder from clogging the line, since capillary action will tend to draw the solder into the joint and thus clog the line. After making a silver solder repair, test the line for leaks and make sure no restrictions exist.

Most leakage trouble occurs in the fuel lines, but leaks may occasionally develop in the fuel tank. These leaks must be eliminated immediately, because of the potential fire hazard.

The principal causes of fuel tank leakage are improper welds and metal fatigue. Metal fatigue is usually the result of inadequate support at the source of trouble; excessive stresses develop in the tank, and cracks result. If leaks occurring at welded joints, or because of metal fatigue, cannot be repaired aboard ship, the repair item must be carried in the Current Ship's Maintenance Project (CSMP) for the next availability period.

Another factor that can limit the fuel supply to such an extent that an engine will not start is clogged fuel filters. As soon as it is known that clogging exists, the filter elements should be replaced. Definite rules for such replacement cannot be established for all engines. Instructions generally state that elements will be used no longer than a specified time and there are reasons why an element may not function properly even for the specified interval.

Filter elements may become clogged because of dirty fuel, too small a filter capacity, failure to drain the filter sump, and failure to use the primary strainer. Usually, clogging is indicated by such symptoms as stoppage of fuel flow, increase in pressure drop across the filter, increase of pressure up-stream of the filter, or excessive accumulation of dirt on the element, observed when the filter is removed for inspection. Symptoms of clogged filters vary in different installations, and each installation should be studied for external symptoms, such as abnormal instrument indications and engine operation. If external indications are not apparent, visual inspection of the element will be necessary, especially if it is known or suspected that dirty fuel is being used.

Fuel filter capacity should at least equal fuel supply pump capacity. A filter with small capacity clogs more rapidly than a larger one, because the space available for dirt accumulation is more limited. There are two standardized sizes of fuel filter elements—large and small. The small element is the same diameter as the large but is only one half as long. This construction permits substitution of two small elements for one large.

The interval of time between element changes can be increased by making use of the drain cocks on a filter sump; removal of dirt through the drain cock will make room for more dirt to collect. If new filter elements are not available for replacement, and the engine must be operated, you can wash some types of totally clogged elements and get limited additional service. This procedure is for emergencies only. An engine must never be operated unless all the fuel is filtered; therefore a "washed filter" is better than none at all. When washing a clogged element, plug the ends preferably with corks. If the ends are not plugged, dirt from the outside of the element will wash into the downstream side of the element, and will be carried into the injection equipment as soon as the engine is started.

Fuel must never flow from the supply tanks to the nozzles without passing through all stages of filtration. *Strainers*, as the primary stage in the fuel filtration system, must be kept in good condition if sufficient fuel is to flow in the system. Most strainers are equipped with a blade mechanism, which is designed to be turned by hand. Do not allow your men to apply a tool or other torque-magnifying device to operate this strainer blade mechanism. If the scraper element cannot be turned readily by hand, the strainer should be disassembled and cleaned. This minor preventive maintenance will prevent breakage of the scraping mechanism.

If the supply of fuel oil to the system is to be maintained in an even and uninterrupted flow, the fuel *transfer pumps* must be functioning properly. These pumps may become inoperative or defective to the point where they fail to discharge sufficient fuel for engine starting. Generally, when a pump fails to operate, some parts have to be replaced or reconditioned. For some types of pump, it is customary to replace the entire unit. However, in the case of worn packing or seals, satisfactory repairs may be made. If plunger-type pumps fail to operate because the valves have become dirty, submerge and clean the pump in a bath of Diesel oil.

Repairs of fuel transfer pumps should be made in accordance with maintenance manuals supplied by the individual pump manufacturers.

## Malfunctioning of the Injection System

The fuel injection system is the most intricate of the systems in a Diesel engine and the troubles which may occur depend upon the system in use. The principal system in use is the *solid injection* system.

Solid injection systems may be divided into two classes: the *jerk pump* system, also called the *individual pump system*, and the common RAIL system. Since an injection system functions to deliver fuel to the cylinder at a high pressure, at the proper time, in the proper quantities, and properly atomized, it is evident that special care and precautions must be taken in making adjustments and repairs.

If a *high-pressure pump* in a fuel injection system becomes inoperative, the engine may fail to start. Information on the troubles, which make a pump inoperative, etc., and the information necessary for overcoming such troubles, requires space than what is available here. Any ship using fuel injection equipment should have on hand copies of the applicable manufacturer's instruction book.

*Unsuitable Fuel Oil* is another factor, which may prevent an engine from starting. As previously mentioned, clean proper stages of filtration is important. Fuel used in diesel engines should meet manufacturers design specifications.

Regardless of the installation or the type of fuel injection system used, maximum energy obtainable from fuel cannot be gained if the timing of an injection system is incorrect. Early or late injection timing may prevent an engine from starting. If the engine does start, it will not perform satisfactorily. Operation will be uneven and vibration will be greater than usual.

If fuel enters a cylinder too early, detonation generally results, causing the gas pressure to rise too rapidly before the piston reaches top dead center. This in turn causes a loss of power and high combustion pressures. Low exhaust temperatures may be an indication that fuel injection is too early.

When fuel is injected too late in the engine cycle, overheating, lowered firing pressure, smoky exhaust, high exhaust temperatures, and loss of power may occur. Correction of an improperly timed injection system should be accomplished by following the instructions given in the appropriate engine manual.

**Insufficient Compression**

Proper compression pressures are essential if a diesel engine is to operate satisfactorily. Insufficient compression may be the reason for an engine failing to start. If low pressure is suspected as being the reason for engine starting trouble, compression should be checked with the appropriate instrument. If the test indicates pressures below standard, complete inspection and correction generally requires some disassembly except in the case of an open indicator valve.

An *inoperative engine governor* may prevent an engine from starting. There are many troubles, which may render a governor inoperative, but those encountered when starting an engine are generally caused by bound control linkage or, if the governor is hydraulic, by low oil level. Regardless of whether the governor is mechanical or hydraulic, binding of the linkage is generally due to parts, which are distorted, misaligned, defective, or dirty. If binding is suspected, linkage and governor parts should be moved and checked by hand. Any undue stiffness or sluggishness in the movement of the linkage should be eliminated.

Low oil level in hydraulic governors may be due to leakage of oil from the governor, or to failure to maintain oil level specified by the instruction manual. Oil leaking from a governor can be traced to a faulty oil seal on the drive shaft or power piston rod, or to a poor gasket seal be- tween parts of the governor case.

The condition of oil seals should be checked if it becomes necessary to add oil too frequently to governors with independent oil supplies. Depending on the point of leakage, oil seal leakage may or may not be visible on external surfaces. There will be no external sign if leakage occurs through the seal around the drive shaft, while leakage through the seal around the power piston will be visible.

Oil seals must be kept clean and pliable, therefore storage must be such that the seals do not become dry and brittle, or dirty. The repair of leaky oil seals requires a replacement. Some of the leakage troubles can be prevented if proper installation and storage instructions for oil seals are followed. Most manufacturers' instruction books give information on the governor installation and trouble shooting.

*Inoperative Overspeed Safety Devices* may be the source of trouble when an engine fails to start. These devices are designed to shut off fuel or air in the event engine speed becomes excessive. However, if an overspeed device is accidently tripped when attempting to start an engine, the shut-off control will prevent starting.

*Inoperative Cold Starting Devices.* If an engine fails to start when the temperature is low, the source of the trouble may be due to defects in the cold starting devices. These devices, generally called *air heaters* or *preheaters*, serve as an external means of heating intake air. This facilitates starting if the outside temperature is low enough to prevent ordinary compression ignition. Two of the air heaters used extensively by the Navy are the electrical type and the flame-priming type.

When an electric air heater fails to operate, check the connections and the filament. Poor electrical connections constitute the difficulty encountered most frequently in this type of air heater. The constant vibration of the engine may cause loose connections, and a poor connection may result if switch contacts become fouled or burned. The filament of this type heater may break because of vibration or extensive use. Engine casualties of this sort may be prevented by conducting proper inspections and following maintenance procedures.

Most troubles encountered in *flame primers,* are related to ignition, oil spray, or the pressure pump. Ignition trouble may occur in the spark coil vibrator, wiring, switch, or spark coil. If vibrator points are in poor condition, they should be cleaned with fine sandpaper or a special point file, and the gap reset. Faulty coils require replacement. It is in the diesel engineer's best interest to make sure that the required inspections and maintenance procedures on electrical component parts or systems of a diesel engine for which you are responsible is being done correctly by qualified personnel.

If the pressure pump of a flame primer is hard to operate, a clogged nozzle may be the source of trouble. A clogged nozzle prevents proper oil spray and may cause the primer to fail. The possibility of clogged nozzles is reduced by properly instructing operating personnel in cleaning the filter screens and nozzles periodically.

The check valves or the plunger piston cups may cause the most common pressure pump failure on a flame primer. Check valves may stick or leak because of dirt. Forcing gasoline through the check valve break foreign debris. Replace piston cups on the pump plunger before they wear out. When replacing a piston cup, make sure the edge is not turned back or torn as the cup enters the pump cylinder. A few drops of lubricating oil on the cup will help with installing them in the pump cylinder.

*Insufficient Cranking Speed* may be the reason for an engine failing to start. With the engine cranking slowly, the compression temperature becomes harder to reach. Low starting air pressure is normally the cause of slow cranking speed and insufficient compression.

Slow cranking speed may also be the result of an increase in the viscosity of the lubricating oil. This trouble occurs when the air temperature is lower than usual. The oil specified for use during normal operation and temperature is not generally suitable for cold climate operation. The oil temperature increase, so does the viscosity and frictional resistance. The greatest friction is in the oil film between the pistons and cylinder walls. To prevent increase in the frictional resistance between the pistons and cylinder walls, engine manufacturers recommend using an oil of lower viscosity for use when the temperature is below a specified point.

During cold weather conditions, heating the cylinder walls can reduce oil viscosity. A method of doing so is by carefully increasing the temperature of the jacket water. Some engines utilize a keep warm pump which is an electric heater built into the cylinder jacket to maintain cylinder wall temperatures. Another cold-weather starting procedure is to admit low-viscosity oil to the lubricating oil pressure pump just before securing the engine. This emits a low-viscosity oil film between the pistons and cylinder walls. In some areas of extremely low temperature, the lightest of recommended oils is too heavy. In such cases, it may be necessary to preheat the oil, or in the absence of preheating facilities, to dilute the oil. If preheating or dilution of lubricating oil is used, adhere to approved prescribed methods only.

Understanding how to keep an engine operating satisfactorily and recognizing the symptoms that might eventually cause major breakdowns is important. In addition, experience is the best teacher when learning how to detect the symptoms of impending problems.

As previously mentioned, the factors that may prevent an engine from starting are: 1) engine cannot be cranked nor barred over, 2) engine cannot be cranked but can be barred over and 3) engine will crank but fails to start. If the engine fail to start or will not be barred over, check for obstructions or seized components, such as a disengaged starting gear, water or oil in the cylinders, or a seized piston or bearing. A malfunctioning starter system is generally why an engine will not crank but still turns over. Keep in mind a number of potential troubles can be source if an engine cranks but fails to start. The fuel and fuel system are the source of a majority of the problems when this occurs. However, the possibility of a defective or inoperative part or assembly and failing to follow the prescribed starting procedures may be the reason for an engine failing to start.

An engine may start without difficulty, but there is still a possibility that several problems may occur. Various problems prevent an engine from starting, which can also occur after the engine starts, and can cause erratic operation.

**Irregular Engine Operation**

The Navy Diesel Engineer must constantly be alert to detect any symptoms that might indicate trouble. Such symptoms may be sudden or abnormal changes in the supply, temperature, or the pressure of lubricating oil or of cooling water. Color and temperature of exhaust are indicators that abnormal conditions exist. Fuel, oil, and water leaks are an indication of possible troubles. Keeping the engine clean makes it easier to spot such leaks. As a Navy Diesel Engineer, you will soon become accustomed to the "normal" sounds and vibrations of a properly operating engine. An abnormal or unexpected change in the pitch or tone of an engine noise, or a change in the magnitude or frequency of a vibration, gives warning that something is wrong. A new sound such as a knock, a drop in the fuel injection pressure, or a misfiring cylinder are other warnings for which the you should be constantly alert during engine operation.

The following discussion on possible troubles, their causes and the corrective action necessary, is general, rather than specific. The information is based on instructions for some of the engines used by the Navy and is typical of most models of engines in use. In a few cases, you mind find that certain models have trouble that only apply to them. For specific information on any particular engine, consult the owner's manual.

**Engine Stalls Frequently or Stops Suddenly**

Factors that may cause an engine to stall include misfiring, air in the fuel system, clogged fuel filters, unsatisfactory operation of fuel injection equipment and incorrect governor action may cause other troubles as well. For example, clogged fuel oil filters and strainers may lead to loss of power, misfires, erratic firing, and low fuel oil pressure. Single engine trouble does not always manifest itself as a single difficulty, but may be the cause of several major difficulties.

Factors that may cause an engine to stall include misfiring, low cooling water temperature, improper application of load, improper timing, and obstruction in the combustion space or in the exhaust system, insufficient intake air, piston seizure, and defective auxiliary drive mechanisms.

**Misfiring**

When an engine misfires or fires erratically, or when one cylinder misfires regularly, the possible troubles can be usually be associated with the fuel or fuel system, worn parts, or the air cleaner or silencer. You already know the adverse effect of air, dirt, and water on engine operation.

Many of the troubles resulting from fuel contamination require overhaul and repair. However, a cylinder may misfire regularly in some systems because of the fuel pump malfunctioning. Some fuel pumps have this type of mechanism that allows fuel supply to the cylinder to be secured in order to measure compression pressures. If a cylinder is misfiring, save yourself time and trouble by first checking for an engaged cut out mechanism, if installed, and disengage it while the engine is running.

## Loss of Compression

A loss of cylinder compression will cause a cylinder to misfire. Sometimes a leaking cylinder head gasket, leaking or sticking cylinder valves, worn pistons, liners, or rings, or by a cracked cylinder head or block will cause compression loss.

In determining what causes a cylinder to misfire, you should follow prescribed procedures given in the appropriate instruction manual. Procedures will vary between engines because of differences in the design of parts and equipment. The differences in various engines, such as the manner in which air enters and leaves an engine will obviously dictate the procedure for tracing a miss firing or low compression cylinder on the different engines.

If loss of compression pressure causes an engine to misfire, check of the compression pressure for each cylinder. Some indicators measure compression as well as firing pressure while the engine is running at full speed. Others check only the compression pressures, with the engine running at a relatively slow speed.

After installing the indicator, operate the engine at the specified rpm and record the cylinder compression pressure. Follow this procedure on each cylinder in turn. The pressure in any one cylinder should not be lower than the specified number of pounds per square inch, nor should the pressure for any one cylinder be excessively lower than the pressures in the other cylinders. Data sheets and manufacturers manuals provide maximum pressure variation permitted between cylinders. A compression leak is indicated when the pressure in one cylinder is considerably lower than that in the other cylinders.

A test indicating a compression leak means some engine disassembly, inspection, and repair. Also, check valve seats and cylinder head gaskets for leaks and inspect the valve stems for sticking. A cylinder head or block may be cracked. If these parts are not the source of trouble, compression is leaking due to insufficient sealing of the piston rings. Sometimes the reason for an engine firing erratically or misfiring is the clogging of air cleaners and silencers.

At specified intervals, clean air cleaners as recommended in engine instruction manuals. A clogged cleaner reduces the intake air, thereby affecting the operation of the engine. Clogged air cleaners may cause not only misfiring or erratic firing, but also such difficulties as hard starting, loss of power, engine smoke, and overheating. Volatile solvents are excellent cleaning agents for cleaning an air cleaner element, but may cause an explosion if the cleaner is not dry before reinstallation.

Air cleaners and filters of the oil bath type are the source of very little trouble if serviced properly. Cleaning directions are generally on the cleaner housing. Usually, on operating hours determine cleaning frequency, but more frequent cleanings may be necessary where unfavorable conditions exist. When filling a cleaned oil bath type cleaner, it is necessary that filling instructions be followed. Most air cleaners of this type have a FULL mark on the oil reservoir. Filling beyond this mark does not increase the efficiency of the unit and may lead to serious trouble. When the oil bath is too full, the intake air may draw oil into the cylinders. This excess oil-air mixture, over which there is no control, may cause an engine to "run away," resulting in serious damage.

**Low Cooling Water Temperature**

If an engine is to operate properly, maintaining the cooling water within specified temperature limits are essential. When cooling water temperature becomes lower than recommended for a diesel engine, this increases ignition lag, causing detonation that results in "rough" operation and may cause an engine to stall. The thermostatic valves that control cooling water temperature operate with a minimum of trouble. Cooling water temperatures above or below the value specified in the instruction manual sometimes indicate that the thermostat is inoperative.

However, high or low cooling water temperature does not always indicate thermostat trouble. The engine load may be insufficient to maintain proper cooling water temperatures, or the temperature gage may be inaccurate or inoperative. Check these items before removing a thermostatic control unit. The thermostatic valves (similar to automotive type thermostats) may fail to function because of a hole in the element. This is the most frequent cause of faulty operation in this type of thermostatic valve. Ensure you handle thermostat elements carefully to avoid puncturing. When a faulty thermostat is suspected, be removed it from the engine and test it according to the procedure below.

1. Obtain an open container or bucket which permits freedom of vision.

2. Heat the water to the temperature at which the thermostat is supposed to start opening. Use the temperature provided in the appropriate instruction manual. Use an accurate thermometer to keep a check on the water temperature. Use a hot plate or a burner as a source of heat. Stir the water frequently to ensure uniform distribution of the heat.

3. Suspend the thermostat in such a manner that operation of the bellows will not be restricted. A wire or string will serve as a satisfactory means of suspension.

4. Immerse the thermostat, and observe its action. Check the thermometer readings carefully to see if the thermostat begins to open at the recommended temperature.

5. Increase the temperature of the water until the specified FULL open temperature is reached. The immersed thermostatic valve should be fully open at this temperature.

Replace the thermostat if there is no movement, or if the temperature at which the thermostat begins to open, or opens fully, is different than temperature specified in instruction manuals.

The older Fulton–Sylphon automatic temperature regulator is relatively trouble free. The unit controls temperatures by positioning a valve to bypass some water around the cooler. This system provides a full flow of the water, with only a portion cooled. In other words, the circulating cooling water is at the proper volume and velocity. This eliminates the possibility of the formation of steam pockets in the system.

Generally, when the automatic temperature regulator fails to maintain cooling water at the proper temperature, it is an indication of improper adjustment. However, the element of the valve may be leaking or some part of the valve may be defective. Failure to follow the proper adjustment procedure is the only cause for improper adjustment of an automatic temperature regulator. Check and follow the proper procedure in the instruction manual issued for the specific equipment.

Adjust the regulator by changing the tension of the spring (which opposes the action of the thermostatic bellows) with a special tool used to turn the adjusting stem knob or wheel. Increasing the spring tension raises the temperature range of the regulator, and decreasing it lowers the temperature range.

When placing a new valve of this type in service, you must be take a number of actions to ensure that the valve stem is proper length and that all scale pointers make accurate indications. Make all adjustments and required maintenance according to the valve manufacturer's instruction book.

## Improper Applications of Load

An engine may stall even though operating properly, if the operator applies the load too suddenly. Always take into consideration load an engine and preventing running at slow speed for an extended period. *Improper Timing* of the fuel injectors may be the cause of an engine stalling.

## Obstructions in the Combustion Space

Pay close attention for any broken valve heads, valve stem locks or keepers that come loose because of a broken valve spring may cause an engine to come to an abrupt stop. Severe damage will result if an engine continues to run when such obstructions are in the combustion chamber, the piston, liner, head, and injection nozzle.

## Obstructions in the Exhaust System

This type of trouble is seldom occurs if proper installation and maintenance procedures are followed. When a part of an engine exhaust system is restricted, there is an increase in the exhaust backpressure. This may cause high exhaust temperatures, loss of power, or even stalling.

The manifolds of an exhaust system are relatively trouble free if related equipment is designed and installed properly. Improper design or installation may result in water backing up into the exhaust manifold. In some installations, silencer design may be the cause of water flowing into the engine. The source of water that may enter the engine must eliminated. This may require replacing some parts of the exhaust system with components of an improved design, or relocating such items as the silencer and piping. An obstruction that causes excessive backpressure in an exhaust system is associated with the silencer or muffler. Inspect exhaust manifolds for water or symptoms of water.

Accumulation of salt or scale in the manifold usually indicates that water has been entering from the silencer. Turbochargers on some engines have been known to seize because of salt water entered the exhaust gas turbine from the silencer. The presence of corrosion or of salt deposits on the engine exhaust valves indicates entry of water into an engine.

If inspection reveals signs of water in an engine or the exhaust manifold, take steps immediately to correct the trouble. Check the unit for proper installation. Install wet type silencers with the proper sizes of piping. If the inlet water piping is too large, too much water may be injected into the silencer.

Excessive accumulation of oil or soot sometime clog dry type silencers. When this occurs, exhaust backpressure increases, causing troubles such as high exhaust temperatures, loss of power, or possibly stalling. A dry type silencer clogged with oil or soot is also subject to fire. Fire, soot, and sparks from the exhaust stack can usually detect this.

An excessive accumulation of oil or soot in a dry type silencer may be due to a number of factors, such as failure to drain the silencer, poor condition of the engine, or improper engine operating conditions. Clean silencers at necessary intervals of oil and soot accumulations. Depending upon engine usage cleaning operation may require frequent inspections and cleaning. For example, an accumulation of soot and oil is more likely to occur during periods of prolonged idling than when the engine is operating under a normal load. Idling periods should be to a minimum.

## Insufficient Intake Air

Insufficient intake air, which may cause an engine to stall or stop, may be due to blower failure or to a clogged air silencer or air filter. Even though all other engine parts function perfectly, efficient engine operation is impossible if the air intake system fails to supply a sufficient quantity of air for complete combustion of the fuel.

## Blower Failure

Damage to the rotor shaft, thrust bearings, turbine blading, nozzle ring, or blower impeller may prevent a centrifugal blower from performing its function as designed.

Damage to the rotor shaft and thrust bearings usually result from insufficient lubrication, an unbalanced rotor, or operation with excessive exhaust temperature.

Centrifugal blower lubrication problems could be the result of failure of the oil pump to prime, low lube oil level, clogged oil passages or oil filter, or a defective relief valve unable to maintain proper lube oil pressure.

If an unbalanced rotor is the cause of shaft or bearing trouble, there will be excessive vibration. Unbalance may be partly due to a damaged turbine wheel blading, or by a damaged blower impeller.

Operating a blower when the exhaust temperature is above the specified maximum safe temperature generally causes severe damage to turbocharger bearings, and other parts. Make every effort to find and eliminate causes of excessive exhaust temperature before damaging the turbocharger.

Turbine blading damage in a centrifugal type blower could be due to operating with an excessive exhaust temperature, by operating at excessive speeds, by bearing failures, by failure to drain the turbine casing, or by the entrance of foreign bodies. Nozzle ring damage may be caused by excessive exhaust gas temperature, foreign bodies, and lose turbine blades.

Damage to an impeller of a centrifugal blower may result from thrust or shaft bearing failure, entrance of foreign bodies, or loosening of the impeller on the shaft.

Since blowers are high-speed units and operate with a very small clearance between parts, it is obvious that minor damage to a part might result in extensive blower damage and failure. All Navy Diesel Engineers must know the proper operating and maintenance procedures. Lack of knowledge will probably result in the need for blower replacement or repair.

Even though there is considerable difference in principle and construction of the positive displacement blower and the axial-flow positive displacement blower, the problems of operation and maintenance are similar. Some of the troubles encountered in a positive displacement type blower are similar to those already mentioned under the discussion of the centrifugal type blowers. However, the source of some troubles may be different because of construction differences. Positive displacement type blowers are equipped with a set of gears to drive and synchronize the rotation of the rotors.

Whether these blowers are driven by a serrated shaft or not, the basic problem in both types of blowers is to maintain the necessary small clearances. Damaged rotors and case can be attributed to the inability to if clearances cannot be maintained which prohibits the blower from perform its function.

Worn gears are one source of trouble in positive displacement type blowers. A certain amount of gear wear is expected, but damage resulting from excessively worn gears indicates improper maintenance procedures. Contaminated lube oil is the primary cause of excessively worn gears. The supply of lubricating oil comes from the engine system. Maintaining the engine system properly, and the oil lines and passageways kept open, no trouble should result from lack of, or contamination of, oil. Replace worn gears as necessary. Knowing if replacement is required is informed by measuring the backlash of the gear set. Record the values of backlash in the machinery history of the engine. This record establishes the baseline rate of wear and estimates the life of the gears. From this record, it can be determined when it will be necessary to replace the gears.

Scored rotor lobes and casing may cause blower failure. Worn gears, improper timing, or foreign matter may cause scoring of blower parts. Any of these troubles may be serious enough to cause contact of the rotors and extensive damage to the blower.

Timing of blower rotors not only involves gear backlash but also the clearances between leading and trailing edges of the rotor lobes and between rotor lobes and casing. Measure the thickness between these parts with thickness gages. If clearances are incorrect, check the backlash of the drive gear first. If the backlash is excessive, replace the gears. Then retime the rotors according to the method outlined in the appropriate manufacturer's instruction book.

Failure of serrated blower shafts may be a result of failure to inspect the parts or improper replacement of parts. When inspecting serrated shafts, be sure that they fit snugly and that wear is not excessive. When serrations of either the shaft or hub have failed for any reason, replace both parts.

**Piston Seizure**

Trouble of this nature may be the cause for an engine stopping suddenly. The piston becomes galled and scuffed. When this occurs, the piston may break or extensive damage to other major engine parts will occur. The principal causes of piston seizure are insufficient clearance, excessive temperatures, or inadequate lubrication.

**Defective Auxiliary Drive Mechanisms**

Such mechanisms, used for the transmission of power within an internal combustion engine, may consist of a *gear train* or a *chain drive*. Defects may occur in mechanisms of this type to cause an engine to stop suddenly. Since most troubles in gear trains or chain drives require some disassembly, this discussion will be limited to such problems.

Gear failure is the principal trouble encountered in gear trains. Engine failure and extensive damage can occur because of a broken or chipped gear. If you hear a metallic clicking noise near gear housing, it is almost a certain indication that a gear tooth has broken.

Gears are most likely to fail because of improper lubrication, corrosion, misalignment, torsional vibration, excessive backlash, wiped bearings and bushings, metal obstructions, or improper manufacturing procedures. Good Navy Diesel Engineers are thoroughly familiar with the construction and operation of gear trains as well as with the prescribed maintenance procedures.

During periodic inspections for scoring, wear, and pitting gear shafts, bushings and bearings, and gear teeth must be inspected. All oil passages, jets, and sprays should be cleaned to ensure proper oil flow. All gear-locking devices must fit tightly, to prevent longitudinal gear movement.

Chains are used in some engines for camshaft and auxiliary drives; in others, chains are used to drive certain auxiliary rotating parts. Troubles encountered in chain drives usually result from wear or breakage. Troubles of this nature is likely to result from improper tension, lack of lubrication, sheared cotter pins or misalignment.

Some doubt may exist as to what the difference is between stalling and stopping. In reality, there is none unless we associate certain troubles with each. Minor adjustments or maintenance can fix most of the problems that cause frequent engine stalling. If these problems are not corrected it is quite possible that the engine can be started, only to stall again. Failure to eliminate these problems may lead to frequent stalling and sudden engine shutdowns. For example, seized piston or bearing is a good example of this type of trouble, which require overhaul to eliminate.

**Engine Will Not Carry Load**

Many of the troubles that can lead to loss of power in an engine may also cause the engine to stop and stall suddenly, or may even prevent it from starting. Insufficient fuel and air supply, improperly positioned fuel rack are possible causes of insufficient power in an engine. Such items as insufficient air, insufficient fuel, and faulty operation of the governor are normally the cause for a lack of engine power. Many of these symptoms are directly related, and the elimination of one may eliminate others. The operator of an internal combustion engine may face major equipment difficulties. Notice the connection with starting failures and with engine stalling and stopping.

**Engine Overspeeds**

When an engine overspeeds, either the governor mechanism or the fuel control linkage, as mentioned before usually causes the trouble. When information on a specific fuel system or speed control system is required, check the manufacturer's instruction book and the special maintenance manuals for the particular equipment. These special manuals are available for the most widely used models of hydraulic governors and overspeed trips, and they contain specific details on testing, adjusting, and repairing. When an engine operates at higher than the desired rpm, one possible cause might be an inaccurate tachometer. If the instrument is faulty, it may indicate a reading lower than the actual rpm. The operator will then be likely to try to bring engine speed up to the rated rpm, and thus overload the engine.

Electrical tachometers are the most likely to register incorrect readings. These devices, usually voltmeters, operate on current transmitted from a small generator. The dials are calibrated to indicate rpm speed directly. If there is reason to think that the tachometer recording is inaccurate, a revolution counter should be used to check the readings. If this check reveals that the tachometer is defective, a spare one should be installed, and the defective tachometer should be removed.

**Engine Hunts or Will Not Secure**

Some troubles that may cause an engine to hunt, or vary its rpm at constant throttle speed, are similar to the troubles that cause an engine to resist stopping efforts. Generally, these two forms of irregular engine operation are caused by troubles originating in the fuel system and speed control system. The speed control system of an internal combustion engine includes those parts designed to maintain the engine speed at some exact value, or between desired limits, regardless of changes in load on the engine. Governors are provided to regulate fuel injection so that the speed of the engine can be controlled as the load is applied. The governor also acts to prevent overspeeding as in the case of rough seas when the propellers leaving the water might suddenly reduce the load. If certain parts of the fuel system or governor fail to function properly, the engine may hunt that is, vary at a constant throttle setting or you may have difficulty in stopping the engine.

Fuel control racks that have become sticky or jammed may cause governing difficulties. If the control rack of a fuel system is not functioning properly, the engine speed may decrease as the load is applied, and race as the load is removed; or the engine may hunt continuously, or only when the load is changed. A sticky or jammed control rack may prevent an engine from responding to changes in throttle setting, and may even prevent securing. Any such condition could be serious in an emergency. It is your job to make every effort possible to prevent their occurrence. Check for a sticky rack by disconnecting the linkage to the governor, and then attempting to move the rack by hand. There should be no apparent resistance to the motion of the rack if the return springs and linkage are disconnected. A stuck control rack may occur because of the plunger sticking in the pump barrel, of dirt in the rack mechanism, of damage to the rack, sleeve, or gear, or of improper assembly of the injector pump. The cause of sticking or jamming must be determined and damaged parts must be replaced. If sticking is due to dirt, a thorough cleaning of all parts will probably correct the trouble. Errors in assembly can be avoided by carefully studying the assembly drawings and instructions.

**Injector Leakage**

Leakage of fuel oil from the injectors may cause an engine to continue to operate when shutdown is attempted. Regardless of the type of fuel system, the results of internal leakage from injection equipment are, in general, somewhat the same. Injector leakage will cause unsatisfactory engine operation because of the excessive amount of fuel entering the cylinder. In addition to the possibility of causing continued operation of the engine when a shutdown is attempted, leakage may cause detonation, crankcase dilution, and smoky exhaust, loss of power, and excessive carbon formation on spray tips of nozzles and other surfaces of the combustion chamber.

Accumulation of lube oil in the intake air passages manifold or air box is another trouble that may prevent stopping an engine. Such an accumulation creates an extremely dangerous condition. Excess oil can be detected by removing inspection plates or covers and examining the air box and manifold. If oil is discovered, it should be removed and the necessary corrective maintenance performed.

The oil may cause an engine to *runaway*. This results from oil being drawn suddenly in large quantities from the manifold or air box into the cylinder of the engine. This oil burns in the cylinder, and the engine governor has no control over the sudden increase of speed. An air box or air manifold *explosion* is also a possibility if excess oil is allowed to accumulate. Some engine manufacturers have provided safety devices to reduce the hazards of such explosions. Excess oil in the air box or manifold of an engine also increases the tendency for carbon formation on liner ports, cylinder valves, and other parts of the combustion chamber. The causes of excessive lube oil accumulation in the air box or manifold will vary depending on the specific engine. Generally, either the accumulation is due to an obstruction in the air box or separator drains. Some engine manufacturers, in an effort to reduce the possibility of crankcase explosions and "runaways," have designed a means of ventilating the crankcase. In some cases, this ventilation is accomplished by a passage between the crankcase and the intake side of the blower. In some engines, an oil separator or air maze is provided in the passage between the crankcase and blower intake. In either type of installation, stoppage of the drains will cause an excessive accumulation of oil. It is essential that drain passages be kept open by being properly cleaned whenever it becomes necessary. Oil may enter the air box or manifold from sources other than crankcase vapors. A defective blower oil seal, a carry-over from an oil type air cleaner, or defective oil piping may be the source of trouble.

Another possible source may be an excessively high oil level in the crankcase. Under this condition, an oil fog is created in some engines by the moving parts. An oil fog may be caused also by excessive clearances in the connecting rod and main journal bearings. In some types of crankcase ventilating systems, the oil fog will be drawn into the blower. When this occurs, an abnormal amount of oil may accumulate in the air box. Removal of the oil will not remove the trouble. The cause of the accumulation must be determined and the necessary maintenance accomplished. In most cases, except for the defective blower oil seal, the corrective action necessary is obvious or has already been discussed.

If a blower oil seal is defective, replacement is the only satisfactory method of correction. When installing new seals, be sure the shafts are not scored and the bearings are in satisfactory condition. Special precautions must be taken during installation to avoid damaging oil seals. Damage to an oil seal during installation is usually not discovered until the blower has been reinstalled and the engine put into operation. Be sure an oil seal gets the necessary lubrication. The oil not only lubricates the seal, reducing the friction, but also carries away any heat that is generated. New oil seals are generally soaked in clean light lube oil before assembly.

**Cylinder Safety Valves Pop Frequently**

On some engines, a cylinder relief (safety valve) is provided for each cylinder. The function of this type valve is to open if the cylinder pressures exceed a safe operating limit. The valve opens or closes a passage leading from the combustion chamber to the outside of the cylinder. The valve face is held against the valve seat by spring pressure. Tension on the spring can be adjusted with an adjusting nut, which is locked when the desired setting is attained. This setting varies with the type of engine and may be found by referring to manufacturers' instruction books.

*Proper Lifting Pressure.* Cylinder relief valves should be set at the specified lifting pressure. Continual lifting (popping) of the valves indicates excessive cylinder pressure or malfunction of the valves, of which should be corrected immediately. Repeated lifting of a relief valve indicates that the engine is being overloaded, the load is being applied improperly, or the engine is too cold. Also, repeated lifting may indicate that the valve spring has become weakened, ignition or fuel injection occurs too early, injector is sticking and leaking, too much fuel is being supplied, or, in the case of air injection engines, that the spray valve air pressure is too high. When frequent popping occurs, it is necessary to stop the engine, and determine and remedy the cause of the trouble. In case of an emergency, the fuel supply may be cut off to the cylinder affected. Relief valves must never be locked closed, except in cases of emergency. When emergency measures are taken, the valves must be repaired or replaced, as necessary, as soon as possible.

*Malfunctioning of the High-Pressure Injection Pump.* When excessive fuel is the cause of frequent safety valve lifting, the trouble may be due to the improper functioning of a high pressure injection pump, a leaky nozzle or spray valve, or a loose fuel cam (if adjustable); or, in some systems such as the common rail, the fuel pressure may be too high.

*Defects in Relief Valves.* A safety valve, which is not operating properly, should be removed, disassembled, cleaned and inspected. Check the valve and valve seat for pitting and excessive wear and the valve spring for possible defective conditions. When a safety valve is removed for any reason, the spring tension must be reset. This procedure varies to some extent, depending upon the valve construction. In some 2-stroke cycle double-acting engines, a supercharge valve is provided in the exhaust system for closing the exhaust port at a specified time. Such valves are operated with a chain drive. If the mechanism becomes inoperative, the exhaust ports may be closed when they should be open and the excessive pressure, which results in the cylinder safety valves to pop. Generally, supercharge valves of this type become inoperative because of failure of the shear pin, which acts as a protective device for the chain drive and valves. Except in emergencies, it is advisable to shut an engine down when troubles cause safety valve popping.

Clogged or Partially Obstructed Exhaust Ports may be the cause of cylinder safety valves lifting. This condition will be of infrequent occurrence if proper maintenance procedures are followed. If it does occur, the resulting in- crease in cylinder pressure may be sufficient to cause safety valve popping. Clogged exhaust ports will also cause overheating of the engine, high exhaust temperatures, and sluggish engine operation.

Clogging of cylinder ports can be avoided by removing carbon deposits at prescribed intervals. Some engine manufacturers make special tools for port cleaning. Round wire, brushes of the proper size are satisfactory for this work. Care must be taken when cleaning cylinder ports to prevent carbon from entering the cylinder—the engine should be barred to such a position that the piston blocks the port.

**Symptoms of Engine Trouble**

In learning to recognize the symptoms, locate the cause of an engine trouble, it will be found that experience is the best teacher. Even though written instructions are essential for efficient trouble shooting, the information usually given serves only as a guide. It is very difficult to describe the sensation you should feel when checking the temperature of a bearing by hand; the specific color of exhaust smoke when pistons and rings are worn excessively; and, for some engines, the sound you will hear if the crankshaft counterweights come loose. The equipment must be actually worked with in order to associate a particular symptom with a trouble. However, written information can save a great deal of time, and can eliminate much trial and error. Use the written instructions and detection of troubles in practical situations will be made much easier.

Symptoms, which indicate that a trouble exists, may be in the form of an unusual noise or instrument indication, of smoke, or of excessive consumption or contamination of the lube oil, fuel, or water.

**Noises**

The unusual noises that may indicate that a trouble exists or is impending may be classified as pounding, knocking, clicking, and rattling. Each type of noise must be associated with certain engine parts or systems, which might be the source of trouble.

*Mechanical Pounding or Knocking.* A mechanical knock or hammering (not to be confused with a fuel knock) may be caused by a loose, excessively worn, or broken engine part. Generally, troubles of this nature require more than routine maintenance.

Detonation is caused by the presence of fuel or lubricating oil in the air charge of the cylinders during the compression stroke. Excessive pressures accompany detonation; if detonation is occurring in one or more cylinders, an engine should be stopped immediately to prevent possible damage.

*Metallic Clicking.* Noises of this type are generally those associated with an improperly functioning valve mechanism or timing gear. If the cylinder or valve mechanism is the source of metallic clicking, the trouble may be due to a loose valve stem and guide, insufficient or excessive valve tappet clearances, a loose cam follower or guide, broken valve springs, or a valve that is stuck open. A clicking in the timing gear usually indicates that there are some dam- aged or broken gear teeth.

*Rattling Noises* are generally due to vibrating loose engine parts. However, an improperly functioning vibration damper, a failed antifriction bearing, or a gear type pump operating without prime should not be overlooked as possible sources of trouble when rattling noises occur. When you hear a noise, first make sure that it is a trouble symptom. Diesel engines have a characteristic sound caused by high compression. This sound is sometimes mistaken for a mechanical knock. Knocks may be detected and located by the using a solid iron screwdriver or bar. In some cases, special instruments may be provided for this purpose.

## Instrument Indications

An engine operator probably relies on the instruments to warn him of impending troubles more than on all of the other troubles symptoms combined (noises, smoke, contamination and excessive consumption of fuel oil, lube oil, and water). Regardless of the instrument type used, the indications are of no value if inaccuracies exist. Be sure an instrument is accurate and operating properly. All instruments must be tested at specified intervals or whenever they are suspected of being inaccurate.

## Smoke

The presence of smoke can be quite useful as an aid in locating some types of trouble, especially if used in conjunction with other trouble symptoms. The color of exhaust smoke also can be used as a guide in trouble shooting. The color of engine exhaust gives a good general indication of engine performance. The exhaust of an efficiently operating engine has little or no color. A dark, smoky exhaust indicates incomplete combustion, and the darker the color the greater the amount of unburned fuel in the exhaust. Incomplete combustion may be due to a number of troubles. Some manufacturers associate a particular type of trouble with the color of the exhaust. The more serious troubles are generally identified with either black or bluish-white exhaust colors.

## Excessive Consumption and Contamination of Lube Oil, Fuel, or Water

Even the most inexperienced operator is aware of engine trouble when excessive consumption of any of the essential liquids takes place. The possible troubles signified by excessive consumption will depend upon the system in question, but leakage is one trouble common to all. Before starting any disassembly, check for leaks in the system in which excessive consumption occurs.

### Trouble Shooting Gasoline Engines

The trouble shooting procedures used for a marine gasoline engine are, in many ways, similar to those for a diesel engine. The main parts and systems of the two type engines are, with but two exceptions, quite similar. They differ principally in the manner of getting fuel and air into the cylinders and in the method of ignition. This section deals primarily with those systems that differ in the gasoline and diesel engines. In addition, troubleshooting information is given on the electrical systems. The amount of electrical troubles can be reduced by checking to see that correct operating and maintenance procedures are followed. Most electrical system troubles develop from improper use, care, or maintenance. The following information is given primarily to aid you in detecting electrical troubles in order that corrective action may be taken. When a gasoline engine fails to start, one of three conditions exists. The engine is not free to turn, the starter does not crank the engine, or the engine is cranked but does not start. If the engine will not turn over, some internal component is probably seized. In this event, it is advisable to make a thorough inspection, which may necessarily include some disassembly.

### Starter Does Not Turn

If the starter fails to turn, the trouble can usually be traced to the battery, connections, switch, or starter motor. Symptoms of battery trouble generally occur before the charge gets too low to perform the required work. Battery failure is normally followed by a gradual decline in the strength of the battery charge. A dead battery may be the result of insufficient charging, damaged plates, or improper starting technique. Personnel that operate equipment using storage batteries and personnel that maintain batteries should cooperate in the proper care and maintenance of storage batteries. Supervisory personnel, such as chiefs and first class petty officers, should assure themselves that all conditions are satisfactory and, if necessary, take such steps as may be required to correct any unsatisfactory conditions.

The voltage regulator should be set so the charging rate of the generator is sufficiently high to keep the starting battery from running down rapidly. It must not be set so as to cause charging at a higher rate than that specified on the battery name plate, or the battery will be damaged. If the generator is delivering current at a low rate between engine starts, the charge withdrawn from the battery during the starting period will be greater than that replaced by the generator during the charging period. This usually occurs when the engine is started and secured at frequent intervals or during cold weather.

Generators used to maintain the charge of starting batteries may become defective. When this occurs, the normal symptoms are a low battery charge when the engine is started and a zero or low ammeter reading when the engine is running.

Batteries must be in good condition to ensure the proper operation of an ignition system. A starter draws a heavy current from the best of batteries. When a battery is weak, the voltage will be insufficient to operate the ignition system satisfactorily. This is because the heavy starting current will drop the voltage of the battery to an extremely low value. Flames and sparks of all kinds should be kept away from the vicinity of storage batteries. A certain amount of hydrogen gas is emitted from batteries at all times. In confined spaces, this gas can form a dangerous explosive mixture. When using tools about a battery, personnel should be careful not to short-circuit the battery terminals. Never use a tool or metal object to make a so-called "test" of a storage battery. Batteries in exposed locations subject to low temperatures should be kept fully charged during cold weather. It is highly recommended that stored batteries be kept in a warm compartment during cold weather.

Batteries will function for a longer period if the prescribed starting instructions of the engine manufacturer are followed. Many of the difficulties that are encountered in starting an engine can be prevented by proper attention to maintenance instructions.

For instance, boat engines are often difficult to start in cold weather; and this fact emphasizes the need of fully charged storage batteries, of proper engine maintenance, and of correct starting procedures.

Electrical Connections are another possible source of trouble if the starter does not turn. All connections must be kept tight and free from corrosion if maximum capacity is to be obtained from the battery. Battery terminals, since they are more vulnerable to corrosion, looseness, and burning, are the principal source of trouble. Corrosion results when electrolyte is exposed to the battery terminals. This contact with the terminals may be a result of carelessness while taking readings or adding water, of loose filler caps, of failure to coat terminals with grease, or of sloshing the electrolyte out when handling the battery. A loose connection, a corroded terminal, or a short circuit may cause burned battery terminals. Burning of terminals usually occurs when an engine is being started. Burning may be detected by smoke, a flash, spattering of molten metal, and a frying sound near the battery. Usually, the starting motor will cease to turn after the occurrence of these symptoms.

Switches, electrical relays, or contactors that are defective or inoperative may be the reason for a starter not turning. Contactors, being subjected to extremely high current, must be maintained in the best possible condition. Either manually or magnetically operated starting contactors, both are designed for short periods of operation. Burned contacts, if not repaired or replaced, usually result in a complete failure of a starting circuit. Contact points become burned and worn because of normal wear. Contact points may also burn out if the operating switch is not closed securely when the engine is being cranked, the current will arc across the contacts and the excessive heat will destroy the surfaces. The operating switch must be held down firmly whether the contactor is of the manual or magnetic type. Low battery voltage is one of the major causes of burned contacts in magnetic contactors. A low voltage will cause faulty operation that allows the contactor to vibrate and arc excessively.

For the most part, trace starter motor troubles to the commutator, brushes, or insulation. If motors are to function properly, they must be kept clean and dry; dirt and moisture makes good commutation impossible. Dirty and fouled starter motors may be caused by failure to replace the cover band, by water leakage, or by excess lubrication. On most starter motors, a cover is provided to protect the commutator and windings. If the operator neglects to replace the cover, or removes it, to provide added ventilation and cooling, dirt and water are sure to damage the equipment. Although lubrication of bearings is essential for proper operation, excessive lubrication may lead to trouble in a starter motor. Excess lubricant in the shaft bearings may leak or be forced past the seal, and foul the insulating material, commutator, and brushes. The lubricant prevents a good electrical contact between the brushes and commutator, causing sparking and heating of the commutator, and burning of the brushes.

Burned brushes are another possible source of trouble if the starter motor is inoperative. Loose brush holders, improper brush spring tension, a brush stuck in the holder, a dirty commutator, improper brush seating surface, and overloading starter, may cause burning.

## Starter Motor Operates But Does Not Crank Engine

If the diesel engineer knows the starter motor is in good operating condition but fails to crank the engine, the problem will usually be found in the drive connection between the motor and the ring gear on the flywheel. Troubles encountered in the drive assembly are usually in the form of broken parts or a slipping clutch (if applicable). A slipping clutch may be the result of the engine not being free to turn, or of the clutch not holding up to its rated capacity.

Even though seldom encountered, a stripped ring gear on the flywheel may be the source of trouble if the starter motor does not turn the engine.

## Engine Cranks but Fails to Start

Starting troubles, their cause and correction, may vary to some degree, depending upon the particular engine. If the prescribed pre-starting and starting procedures are followed and a gasoline engine fails to start, the source of trouble will probably be improper priming or choking, a lack of fuel at various points in the system, or a lack of spark at the spark plugs.

*Improper Priming* may be classified as either under-priming or over-priming. Priming instructions differ, depending upon the engine. Information on priming is also applicable to engines equipped with chokes. From your experience, you should be well aware that a warm engine should never be primed. Some engines may require no priming except when starting under cold weather conditions.

On some installations, under priming can be checked by the feel of the primer pump as it is operated. On other installations, under-priming may be due to insufficient use of the choke. Overpriming results in a flooded engine and makes starting difficult. Overpriming is also undesirable because excess gasoline condenses in the intake manifolds then flowing down into the cylinders, washing away the lubricating oil film, and causing the pistons or rings to stick. Flooding can be determined by removing and inspecting a spark plug. A wet plug indicates flooding.

Deflooding or drying out procedures must be accomplished according to prescribed instructions. Some instructions specify that the ignition switch must be ON, while others state the switch must be OFF; therefore, it is important that appropriate engine manufacturer's instructions be followed.

*Improper Carburetion* may be the source of trouble if a gasoline engine fails to start. On some engines, a check of the fuel pressure gage will indicate whether lack of fuel is the cause. If the gage shows the prescribed pressure, the trouble is not lack of fuel. If the gage shows little or no fuel pressure, then the various parts of the fuel delivery system must be checked to locate the fault. In some installations, whether the trouble is in the gage or in the fuel system can be determined by the following procedure: (1) remove the carburetor plug next to the fuel pressure gage connection, and (2) use a suitable container to catch the gasoline and operate the pump used to build up starting fuel pressure. If fuel is reaching the carburetor, gasoline will spurt out of the open plughole. This indicates that the gage is inoperative. If no fuel flows from the plug opening, the trouble is probably in the fuel system somewhere between the fuel tank and carburetor. Even though not all installations are equipped with a fuel pressure gage, the procedure for checking the fuel system will be much the same. If a wobble pump is installed to build up starting fuel pressure, using feel and sound operation can be determined. If the pump feels or sounds dry, then the trouble is between the pump and supply tanks. The trouble might be caused by a clogged fuel line strainer, or by an air leak in the line. If the wobble pump is pumping, then the trouble may exist in the line to the engine fuel pump or in the engine fuel pump itself.

Fuel lines should be checked for cracks, dents, loose connections, sharp bends, and clogging. The fuel line can be removed at the pump and air used to determine if the line is open.

Fuel pumps should be checked for leaks at the pump gaskets or in the fuel line connections. Fuel pump filters or sediment bowl screens should be checked for restrictions. The bypass valve of a vane type pump should be checked for operation. If defective, the fuel pump will have to be replaced. In diaphragm type fuel pumps, the filter bowl gasket, the diaphragm, or the valves may be the source of trouble. Air leaks can be checked by submerging the dis- charge end of the fuel line in gasoline and looking for air bubbles while cranking the engine. If the engine will run, a leaky diaphragm can be detected by gasoline leakage out of the air vent. Carburetor trouble may be the reason for fuel not reaching the cylinders. This can be checked by removing the spark plugs and looking for moisture. If there is no trace of gasoline in the cylinders, the carburetor may be out of adjustment, the float level may be too low, or the jets may be clogged. If the fuel level in the carburetor float bowl is low, the float valve is probably stuck on the seat. If the fuel level in the float bowl is correct, yet no fuel is delivered to the carburetor throat, the carburetor will have to be removed, disassembled, and cleaned.

*Faulty Ignition System Parts* may be the source of starting difficulties, if a check of the fuel system reveals its operation is satisfactory. Two kinds of ignition systems—the *magneto type* and the *battery type* may be encountered. Even though the parts of these systems differ in some respects, their function is the same; namely, to produce a spark in each cylinder of the engine at exactly the proper time in relation to the position of the pistons and the crankshaft. In addition, the system is designed so that the sparks in all cylinders follow each other in proper sequence. The inspection, maintenance, and repair of electrical equipment is generally the responsibility of the electrical engineers. Even so, the diesel engineers should be thoroughly familiar with the component parts of electrical systems on machinery for which he is responsible. In unusual circumstances, the diesel engineers be required to care for and maintain electrical equipment; in such cases, he should follow the information given in the manufacturer's instruction book.

## Engine Fails to Stop

If a gasoline engine fails to stop when the ignition switch is turned to the OFF position, either a faulty ignition circuit or an overheated engine will cause the trouble. In a magneto type ignition system, an open *ground connection* may result in an engine running after the ignition switch is turned off. When a magneto ground connection is open, the magneto will continue to produce sparks as long as the magneto armature magnets rotate, and the engine will continue to run. In other words, when the magneto ignition switch contacts points are closed, the ignition should be shut off. This is not true of the booster coil circuit of a magneto type system, nor of the usual battery type ignition system. In these systems, an open ground or open switch points prevent current flow. If the switch of a battery type ignition system fails to stop an engine, it is probable that the contact switch points have remained closed. If the ignition switch and the circuit are in good condition, failure to stop may be caused by overheating. If the engine is overheated, normal compression temperature may become sufficiently high to ignite the fuel mixture even though no spark is being produced in the cylinders. When this occurs in a gasoline engine, the engine is in reality operating on the Diesel principle.

Normally, the symptoms of overheating will be detected before the temperature gets too high. The causes of overheating in a gasoline engine are much the same as those for a diesel engine. Other troubles—and their symptoms, causes, and correction——which may occur in a gasoline engine are similar to those found in a diesel engine. Gasoline and diesel engines experiencing a loss of rpm, irregular operation, unusual noises, abnormal instrument indications, and excessive consumption or contamination of the lube oil, fuel, or water, usually remedied with the same corrective actions. Of course, there are always exceptions, so it is best to use the appropriate engine instruction manual.

## Summary

This chapter provided basic information regarding the troubles encountered when an engine does not perform properly, and to interpret the symptoms and warnings of impending trouble. A section is devoted to the trouble shooting of gasoline engines. You should know the reasons why an engine stalls frequently or stops suddenly. You should have a general idea of how to trace a misfiring cylinder and how to check cylinder compression pressure.

You should know why an engine overspeeds or hunts, as well as why it will not carry a load or shut off. Do not overlook the importance of cylinder relief valves, why they may pop frequently, and what actions you can do to prevent this reoccurring problem. Only practical experience will teach you the specific details involved in maintaining any one installation.

The necessity of practical experience cannot be overemphasized when you are learning to recognize the symptoms of troubles. This is especially true with respect to the abnormal noises that may occur during engine operation. This chapter gives only a general classification of these noises. This classification should serve as a general guide when you attempt to identify an abnormal noise. You should be familiar with the trouble symptoms indicated by the engine instruments.

One thing you must keep in mind is that the reliability of instrument indications depends upon the accuracy of the instrument. You should be familiar with the correlation of certain instrument indications and know how such correlation is helpful in locating some types of trouble. You should have a general idea as to how smoke may serve as an aid in locating some types of trouble; also, how the color of exhaust smoke may serve as a guide in trouble shooting.

You should be able to identify the causes of excessive consumption or contamination of lube oil, fuel, or water. Keep in mind that leakage, as one cause of excessive consumption, is common to all three systems and should be checked before starting disassembly. Even though contamination of the liquids may be harder to detect than excessive consumption, symptoms will be apparent on close examination.

You must know these symptoms and be constantly on the alert for any troubles, which cause contamination. When trouble shooting an engine, keep in mind that a trouble may be evidenced by more than one symptom. Learn to associate as many symptoms as possible with a particular trouble.

# INDEX